W0016609

SPECIAL PUBLICATIONS
THE MUSEUM
TEXAS TECH UNIVERSITY

**Contributions in Mammalogy in Honor of
Robert L. Packard**

Edited by
Robert E. Martin and Brian R. Chapman

No. 22   September 1984

TEXAS TECH UNIVERSITY

Lauro F. Cavazos, President

Regents.—B. Joe Pevehouse (Chairman), John E. Birdwell, J. Fred Bucy, Jerry Ford, Rex Fuller, Nathan C. Galloway, Larry D. Johnson, Wesley W. Masters, and Anne W. Sowell.

Academic Publications Policy Committee.—John R. Darling (Chairman), Robert J. Baker, Weldon E. Beckner, Dilford C. Carter, E. Dale Cluff, David L. Higdon, Clyde Jones, J. Knox Jones, Jr., John L. Kice, C. Reed Richardson, Charles W. Sargent, and Idris R. Traylor.

The Museum
Special Publications No. 22
234 pp.
28 September 1984
Paper, $25.00
Cloth, $50.00

Special Publications of The Museum are numbered serially, paged separately, and published on an irregular basis under the auspices of the Vice President for Academic Affairs and Research and in cooperation with the International Center for Arid and Semi-Arid Land Studies. Copies may be purchased from Texas Tech Press, Sales Office, Texas Tech University, Lubbock, Texas 79409, U.S.A. Institutions interested in exchanging publications should address the Exchange Librarian at Texas Tech University.

ISSN 0149-1768
ISBN 0-89672-123-X (paper)
ISBN 0-89672-124-8 (cloth)

Texas Tech Press, Lubbock, Texas

1984

CONTENTS

PREFACE

Shortly after the passing of Dr. Robert L. Packard, several of his former students discussed the possibility of having a volume of contributed papers in his honor. From a preliminary pool of fourteen former students that agreed to prepare papers, ten submitted papers that were sent out for peer review, accepted for publication, and subsequently incorporated into this volume. The foreword and contents reveal that Bob Packard's students share a diversity of research interests with an underlying appreciation for mammals as subjects for study. Many of us acquired this interest at an early stage in our careers, but Bob Packard strongly influenced, by his love of mammals, our continued interest in mammalian research.

The editors of this volume appreciate the efforts of the many professionals that carefully reviewed the papers and the cooperation and confidence given us by the former students during this project. We also wish to acknowledge the support of Texas Tech University, particularly, Dr. J. Knox Jones, Jr., Vice President for Research and Graduate Studies, Dr. Dilford C. Carter, Executive Director of Academic Publications and Director of Texas Tech Press, and Dr. Robert J. Baker, Chairman, Museum Publications Committee, for seeing this project to completion.

Robert E. Martin
Brian R. Chapman

Robert L Packard

ROBERT LEWIS PACKARD—1928-1979

J. Knox Jones, Jr., and Robert J. Baker

Robert Lewis Packard, known to his friends as "Bob" and to intimates as "Freddy," passed away in Lubbock, Texas, on 8 April 1979, after a lengthy bout with cancer. He was born in Lincoln, Nebraska, on 10 August 1928, one of two children of Wayne and Josephine Packard. After completion of his public school education in 1947, he entered the University of Nebraska, from which he graduated with a double major in Botany and Zoology in 1951. Following two years of military service as an officer in the Ordinance Corps of the U.S. Army, Packard undertook graduate training at the University of Kansas, where he earned the M.A. degree in 1955 and the Ph.D. degree in 1960, the former under the direction of Rollin H. Baker and the latter under the guidance of E. Raymond Hall.

Early in his college career, while still an undergraduate, what was to become an abiding interest in vertebrate zoology was evident, with an initial interest in birds as indicated by his first publications. Gradually, Packard shifted his major emphasis to mammals and most of his 61 published contributions related to that group. After completion of his doctoral work, he was appointed Assistant Professor of Biology at Stephen F. Austin State College in Texas, where he taught from 1959 to 1962. In the latter year, he moved to Texas Technological College (soon to become Texas Tech University), with which he was associated until his death. There, he was Assistant Professor of Biology from 1962 to 1965, Associate Professor from 1965 to 1967, and Professor thereafter. He also served that institution as Assistant Dean of the Graduate School (1967-68), Coordinator of Research for The Museum (1971-75), and Director of the Junction Center campus (1975-79).

As a child, Dr. Packard was stricken with a rare blood disease, hypoplastic anemia, which caused him to miss a year of grammar school, plagued him to varying degrees throughout his life, and probably contributed to his untimely death. Despite this affliction, he led an active life, participating in athletics in high school, where he earned a varsity letter in golf, and in a wide variety of activities, including extended zoological field work, as an adult. His accomplishments are all the more appreciated in light of his medical history.

On 27 December 1950, Packard married Patricia Joann Croskary, who survives him. To this union were born two daughters, Lisa Ann in 1954 and Lori Sue in 1960 (a first child, a son, was stillborn in 1952). Sadly, the girls were diagnosed as having the same disease as their father but, unlike him, were unable to combat it through childhood. Lisa died in 1965 and Lori in 1968. This personal tragedy, which could have destroyed a lesser man, was borne by Bob, and by his wife, courageously and with determination. Only those who knew them well were acquainted with their deeply-felt emotions.

Bob Packard was an excellent teacher, both in the classroom and in the field, and was officially recognized by Texas Tech for his outstanding contributions to undergraduate education at that institution. He had excellent rapport with students and was regarded by many of them as a kind of "academic father." It seems quite likely that his appreciation for young people and his identification with them was partly a response to the loss of his own children. In any event, he will be remembered by many as an inspiring instructor and a warm human being.

At the graduate level, too, Packard was a successful mentor of students. He directed 23 master's theses (four at Stephen F. Austin State and 19 at Texas Tech) and guided eight candidates to completion of the Ph.D. degree (all at Tech), including the first awarded in the Department of Biological Sciences at that institution. These students and their academic progeny will serve as a permanent legacy to Bob Packard's work as a graduate advisor and director of graduate research. Those who received the master's degree as his students were: Peter V. August (1976); J. Hoyt Bowers (1962); Brian R. Chapman (1970); James L. Crain (1962); Paul G. Desha (1964); Michael H. Droge (1976); Robert B. Drotman (1967); Herschel W. Garner (1965); William B. Grabowski (1964); Michael J. Harvey (1962); Duane Ikenberry (1964); Gerald L. Johnson (1972); Frank W. Judd (1968); Michael C. Krenz (1977); Robert C. McReynolds (1968); Joyce W. Mize (1969); James R. Phillips (1962); Paul R. Ramsey (1969); Jimmy D. Roberts (1969); Frank Schitoskey, Jr. (1967); David J. Schmidly (1968); Timothy L. Tandy (1978); and Daniel R. Womochel (1968). Doctorates completed under Packard's tutelage included: John W. Clarke (1979); Walter H. Conley (1971); Herschel W. Garner (1970); Graham C. Hickman (1974); Robert E. Martin (1974); Kenneth G. Matocha (1975); James B. Montgomery (1979); and Robert W. Wiley (1972).

In addition to formal instruction and directing graduate studies, Bob Packard had considerable impact on several other academic areas of Texas Tech University. He was the first mammalogist at the University and it was primarily as a result of his efforts that one of us (Baker) came to Tech. He was the first to envision that institution as building a quality research program in mammalogy, and he continually worked toward that goal. Today, there are seven professional mammalogists at Texas Tech, to say nothing of those in allied fields with an interest in mammals.

He started the collection of Recent mammals, now housed in The Museum, in 1962 and he served as Curator of Mammals from 1962-71. This collection now contains more than 40,000 catalogued specimens and is among the largest in the United States.

Under his leadership as Director of the Junction Center campus (1975-79), that satellite operation became an active center of education, including summer courses in such areas as education, art, biology, and agriculture. The largest educational building at Junction now bears his name.

Packard was associated with many scientific and learned societies. He joined the American Society of Mammalogists in 1949 while still in undergraduate school, and served the Society in a number of official capacities: as an elected Director for a total of five terms and 11 years until his death (1968-75, 1976-79); as a member of standing committees on Conservation of Land Mammals, 1961-63, Membership (as chairman), 1968-75, and Program, 1975-78; and as Chairman of the Local Committee for the 56th Annual Meeting in Lubbock, Texas, in 1976. He also served, among others, the Texas Academy of Science as a vice president, and the Southwestern Association of Naturalists as journal editor, permanent secretary, president-elect, and as president in 1972-73. These activities, along with his published contributions to the discipline of mammalogy, which spanned nearly three decades and several subdisciplines of the field (and of which a listing follows), are ample testimony to Packard's direct impact on the science.

Packard's name is associated with the original descriptions of six taxa of Recent mammals. He described the following subspecies: *Baiomys musculus handleyi; Baiomys musculus pullus; Baiomys taylori canutus; Baiomys taylori fuliginatus; Ochrotomys nuttalli floridanus*; and *Ochrotomys nuttalli lisae*. He was joint describer of a fossil species, *Baiomys intermedius* Packard and Alvarez, 1965, and of a new name for a whale, *Feresa occulta* Jones and Packard, 1956 (the previous name for which was preoccupied). One mammal, *Antrozous pallidus packardi* Martin and Schmidly, 1982, has been named in his honor.

The way a man faces death is a reflection of his character and the final months of Bob Packard's life were extremely difficult, with progressive cancer despite harsh chemotherapy. My (Baker) personal conversations with him led me to believe that he understood that death was imminent but that he considered it important not to let this condition dilute his activities in his last months of life. During that time, I never saw him exhibit self-pity or depression. In fact, he interacted with his colleagues in such a positive manner that many commented that his death had caught them by surprise. Clearly, his final days reflected the strength of Bob Packard's character.

Finally, another personal note (Jones). Bob Packard and I were closely associated for more than 40 years, having grown up but a few blocks apart in Lincoln, Nebraska, and having attended together junior and senior high school, the University of Nebraska, and (save for separation during military service) Graduate School at the University of Kansas. He was a good guy, a sound scientist, an outstanding educator, and a true friend.

Parts of the foregoing obituary appeared in the *Journal of Mammalogy* (62:855-859, 1981) and the West Texas Museum Association's *The Museum Digest* (May-September, 1980).

BIBLIOGRAPHY OF ROBERT L. PACKARD

1950

1. Notes on the nesting of the black crowned night heron at the Valentine National Wildlife Refuge. Nebraska Bird Rev., 13:63-64.

1951

2. Lead poisoning of waterfowl in Lancaster County, Nebraska. Proc. Nebraska Acad. Sci., p. 31.

1954

3. Notes on the defensive behavior of gray and fox squirrels while moving their young. Trans. Kansas Acad. Sci., 57:471-472.

1955

4. Great horned owl attacking squirrel nests. Wilson Bull., 66:272.
5. The coyote on a natural area in northeastern Kansas. Trans. Kansas Acad. Sci., 58:211-221 (with H. S. Fitch).
6. Occurrence of the mink, west of the hundredth meridian, in Kansas. Trans. Kansas Acad. Sci., 58:222-224 (with H. J. Stains).
7. Release, dispersal, and reproduction of fallow deer in Nebraska. J. Mamm., 36:471-473.
8. Pileated woodpecker and American woodcock in southeastern Kansas. Kansas Ornith. Soc. Bull., 6:15.

1956

9. The tree squirrels of Kansas: ecology and economic importance. Misc. Publ. Mus. Nat. Hist., Univ. Kansas, 11:1-67.
10. Feresa intermedia (Gray) preoccupied. Proc. Biol. Soc. Washington, 69:167 (with J. K. Jones, Jr.).

1957

11. Vernacular names for North American mammals north of Mexico. Misc. Publ. Mus. Nat. Hist., Univ. Kansas, 14:1-16 (with E. R. Hall, S. Anderson, and J. K. Jones, Jr.).
12. Broad-winged hawk in Coahuila. Wilson Bull., 69:370.
13. *Myotis keenii septentrionalis* in South Dakota. J. Mamm., 39:150 (with J. K. Jones, Jr.).
14. Carnivorous behavior in the Mexican ground squirrel. J. Mamm., 39:154.
15. The taxonomic status of Peromyscus allex Osgood. Proc. Biol. Soc. Washington, 71:17-19.
16. New subspecies of the rodent Baiomys from Central America. Univ. Kansas Publ., Mus. Nat. Hist., 9:399-404.

1959

17. First record of the pygmy mouse in New Mexico. J. Mamm., 40:146.

1960

18. Speciation and evolution of the pygmy mice, genus Baiomys (abstract in programme for final examination for the degree of Doctor of Philosophy). Publ. Univ. Kansas Grad. School, 4 pp.
19. Speciation and evolution of the pygmy mice, genus Baiomys. Univ. Kansas Publ., Mus. Nat. Hist., 23:579-670.

1961

20. Notes of some amphibians and reptiles from eastern Texas. Southwestern Nat., 6:105-107 (with R. G. Webb).
21. Additional records of mammals from eastern Texas. Southwestern Nat., 6:193-195.

1963

22. Distribution of the black-tailed jackrabbit in eastern Texas. Texas J. Sci., 15:107-110.
23. Small rodents as consumers of pine seed in East Texas uplands. J. Forestry, 61:523-526 (with G. K. Stephenson and P. D. Goodrum).

1964

24. Arboreal nests of the golden mouse in eastern Texas. J. Mamm., 45:369-374 (with H. Garner).
25. Records of some mammals from the Texas High Plains. Texas J. Sci., 16:387-390 (with H. Garner).

1965

26. Range extension of the hooded skunk in Texas and Mexico. J. Mamm., 46:102.
27. Description of a new species of fossil *Baiomys* from the Pleistocene of central Mexico. Acta Zool. Mexicana, 7(4):1-4 (with T. Alvarez).
28. Geographic variation and evolution of high plains fox squirrel populations. Year book Amer. Philos. Soc. (for 1964), pp. 288-289.

1966

29. *Myotis austroriparius* in Texas. J. Mamm., 47:128.
30. Notes on mammals from Washington Parish, Louisiana J. Mamm., 47:323-325 (with J. L. Crain).

1967

31. Octodontoid, bathyergoid, and ctenodactyloid rodents. Pp. 273-290, *in* Recent mammals of the world (S. Anderson and J. K. Jones, Jr., eds.). Ronald Press Co., 453 pp.
32. Cotton rats fail to damage southern pine seedlings. J. Forestry, 65:495-496 (with M. K. Harvey).
33. Two noteworthy records of bats from Chihuahua. Southwestern Nat., 12:332 (with F. W. Judd).
34. Swimming ability in pocket mice. Southwestern Nat., 12:480-482 (with D. J. Schmidly).

1968

35. An ecological study of the fulvous harvest mouse in eastern Texas. Amer. Midland Nat., 79:68-88.
36. Comments on some mammals from western Texas. J. Mamm., 49:535-538 (with F. W. Judd).

1969

37. Golden eagle-livestock relationships: a survey. Spec. Rept. Internat. Center Arid and Semi-arid Land Studies, Texas Tech Univ., 20:1-82 + 40 pp. appendix (with E. G. Bolen).
38. Taxonomic review of the golden mouse, *Ochrotomys nuttalli*. Pp. 373-406, *in* Contributions to mammalogy—a volume honoring E. Raymond Hall (J. K. Jones, Jr., ed.). Misc. Publ. Mus. Nat. Hist., Univ. Kansas, 51:1-428.

1970

39. Distributional notes on some foxes from western Texas and eastern New Mexico. Southwestern Nat., 14:450-451 (with J. H. Bowers).

1971

40. Research programs and their relevance to teaching. Pp. 26-29, *in* Higher education for the Texas agricultural industry (G. W. Thomas and W. L. Cave, eds.). Coordinating Board, Texas College and University System, Austin, 36 pp.
41. Bats of Texas caves. Pp. 122-132, *in* Natural history of Texas caves (E. L. Lundelius and B. H. Slaughter, eds.). Gulf Natural History, Dallas, Texas (with T. R. Mollhagen).
42. Small mammal survey on the Jornada and Pantex sites. U.S. IBP Grassland Biome Tech. Rept., 114:1-48.
43. (Review of) Mammals of Grand Canyon. Southwestern Nat., 16:221-222.

1972

44. Prey remains in golden eagle nests: Texas and New Mexico. J. Wildl. Mgmt., 36:784-792 (with T. R. Mollhagen and R. W. Wiley).
45. The Guinea pig. Encyclopaedia Americana, 13:558-559.
46. Gundi. Encyclopaedia Americana, 13:614-615.
47. Hutia. Encyclopaedia Americana, 14:620.
48. The pig family. Encyclopaedia Americana, 22:780.
49. Small mammal studies on Jornada and Pantex sites, 1970-1971. U.S. IBP Grassland Biome Tech. Rept., 188:1-81.

1973

50. Comments on movements, home range and ecology of the Texas kangaroo rat, *Dipodomys elator* Merriam. J. Mamm., 54:957-962 (with J. D. Roberts).
51. (Review of) Pleistocene and Recent environments of the Central Great Plains. J. Mamm., 54:1022-1023.

1974

52. Observations on the behavior of the Texas kangaroo rat, *Dipodomys elator*. Mammalia, 37:680-682 (with J. D. Roberts).
53. Electrophoretic analysis of Peromyscus comanche Blair, with comments on its systematic status. Occas. Papers Mus., Texas Tech Univ., 24:1-16 (with G. L. Johnson).
54. An ecological study of Merriam's pocket mouse in southeastern Texas. Southwestern Nat., 19:281-291 (with B. R. Chapman).

1975

55. Small mammal studies on Jornada and Pantex sites, 1972. U.S. IBP Grassland Biome Tech. Rept., 277:1-23.

1977

56. Effects of herbivores on seed usage, dispersal, and reproduction with particular reference to mammals. Pp. 211-226, *in* The impact of herbivores on arid and semi-arid rangelands. Proc. 2nd United States/Australia Rangeland Panel (Adelaide, 1972). Australian Rangeland Soc. Publ., Perth, 376 pp.
57. Ochrotomys nuttalli. Mamm. Species, 75:1-6 (with D. W. Linzey).
58. Mammals of the southern Chihuahuan Desert: an inventory. Pp. 141-153, *in* Transactions of the symposium on the biological resources of the Chihuahuan Desert region, United

States and Mexico (R. H. Wauer and D. H. Riskind, eds.). Trans. Proc. Ser. Nat'l. Park Serv., 3:1-658.

1978

59. Baiomys musculus. Mamm. Species, 102:1-3 (with J. B. Montgomery, Jr.).

1979

60. Laboratory manual to vertebrate natural history. Texas Tech Univ., Lubbock, 72 pp.
61. Demographic patterns of small mammals: a possible use in impact assessment. Pp. 333-340, *in* Biological investigations in the Guadalupe Mountains National Park, Texas (H. H. Genoways and R. J. Baker, eds.). Proc. Trans. Ser. Nat'l. Park Serv., 4:1-410 (with P. V. August, J. W. Clarke, and M. H. McGaugh).

MAMMALS OF THE SAN CARLOS MOUNTAINS OF TAMAULIPAS, MEXICO

DAVID J. SCHMIDLY AND FRED S. HENDRICKS

The San Carlos Mountains are an isolated physiographic unit of about 900 square miles (233,100 hectares) rising from the Gulf Coastal Plain midway between the eastern front of the Sierra Madre and the Gulf of Mexico in west-central Tamaulipas (Fig. 1). They rise to an elevation of about 6000 feet (1830 meters) above sea level and are separated by a narrow extent of desert plain from the more extensive Sierra Madre Oriental to the west, which rises to more than 10,000 feet (3050 meters). The San Carlos Mountains are at the southern edge of the Tamaulipan Biotic Province (Dice, 1943). Rainfall estimates range from about 20 inches (50 centimeters) annually in the thorn scrub surrounding the mountains to about 40 inches (100 centimeters) at higher elevations.

A survey of mammals in the San Carlos Mountains was conducted by Lee R. Dice between 22 June and 30 August 1930 as part of a major investigation of the geology and biology of the San Carlos region. The results of that investigation, under the direction of Dr. Louis B. Kellum of the University of Michigan, were published in 1937 in a volume that included Dice's account of the mammals and ecological communities of the mountains. Dice's field notes and specimens are in the Museum of Zoology, University of Michigan. He worked in the vicinity of the following localities: San José Ranch, La Vegonia Mine, Tamaulipeca Ranch, Marmolejo, El Mulato Ranch, and El Milagro Ranch. Dice reported 28 species of mammals inhabiting the San Carlos Mountains and listed four additional species that possibly occurred there. Field parties from the University of Kansas and the University of Illinois made brief visits to the San Carlos area in August 1961 and June 1969, respectively, but they did not make extensive collections of mammals.

Our primary objectives were to determine the present mammalian fauna of the San Carlos Mountains, their geographic and ecologic ranges, and their systematic status. In addition, certain life history observations have been recorded. Three previous papers (Schmidly *et al.*, 1973; Yates *et al.*, 1976; and Baumgardner *et al.*, 1977) detailed some of the significant distributional records of bats obtained during our work in the area.

METHODS AND MATERIALS

We made six major field expeditions to the San Carlos Mountains on the following dates: 4-5 March 1972; 4-15 January 1975; 15-20 March 1975; 31 December 1975-13 January 1976; 21-26 May 1976; and 15-25 May 1977.

TABLE 1.—*Collecting localities visited during our survey of the mammals of the San Carlos Mountains.*

Station	Locality	Longitude	Latitude	Habitat type
1	0.6 mi. SW Rancho Carricitos	99°01′27″	24°35′34″	Deciduous Thickets
2	0.4 mi. W Rancho Carricitos	99°01′41″	24°36′02″	Deciduous Thickets
3	0.7 mi. W Rancho Carricitos	99°01′54″	24°36′02″	Deciduous Thickets
4	1.0 mi. WNW Rancho Carricitos	99°02′05″	24°36′23″	Deciduous Thickets
5	2.2 mi. W Rancho Carricitos	99°03′19″	24°35′43″	Pine-Oak Forest
6	3.4 mi. NNE San Carlos	98°54′54″	24°37′26″	Pine-Oak Forest
7	2.9 mi. W Marmolejo	99°02′59″	24°37′27″	Pine-Oak Forest
8	2.7 mi. W Marmolejo	99°02′47″	24°37′20″	Pine-Oak Forest
9	2.2 mi. W Rancho Carricitos	99°03′19″	24°35′43″	Pine-Oak Forest
10	1.5 mi. WSW Rancho Carricitos	99°02′35″	24°35′46″	Pine-Oak Forest
11	2.8 mi. WSW Rancho Carricitos	99°02′36″	24°35′45″	Pine-Oak Forest
12	2.2 mi. WSW Rancho Carricitos	99°03′39″	24°35′13″	Pine-Oak Forest
13	4.9 mi. SSE San Carlos	98°55′04″	24°32′13″	Riparian Forest
14	1.0 mi. WSW Rancho Carricitos	99°02′09″	24°35′46″	Riparian Forest
15	0.3 mi. SW Rancho Carricitos	99°01′20″	24°35′55″	Riparian Forest
16	0.3-0.5 mi. SW Rancho Carricitos	99°01′20″	24°35′48″	Riparian Forest
17	0.5 mi. SW Rancho Carricitos	99°01′24″	24°35′45″	Riparian Forest
18	0.6-1.0 mi. WSW Rancho Carricitos	99°01′53″	24°35′45″	Riparian Forest
19	1.1 mi. WSW Rancho Carricitos	99°02′13″	24°35′45″	Riparian Forest
20	0.4 mi. NW Rancho Carricitos	99°01′28″	24°36′16″	Thorn Woodland
21	0.3 mi. W Rancho Carricitos	99°01′31″	24°36′00″	Thorn Woodland
22	0.9 WSW Rancho Carricitos	99°02′02″	24°35′47″	Thorn Woodland
23	2.6 mi. WNW San Carlos	98°58′27″	24°35′47″	Thorn Woodland
24	Tinaja	99°00′13″	24°36′16″	Thorn Woodland
25	0.7 mi. E Tinaja	98°59′34″	24°36′18″	Thorn Woodland
26	2.2 mi. SE San Carlos	98°55′09″	24°33′59″	Xeric Thorn Scrub
27	4.3 mi. SE San Carlos	98°54′12″	24°32′08″	Xeric Thorn Scrub
28	0.6 mi. W San Carlos	98°57′03″	24°35′04″	Xeric Thorn Scrub
29	8.5 mi. SSE San Carlos (= Rancho San Francisco)	98°52′54″	24°29′41″	Xeric Thorn Scrub

Camps were established in order to collect the mammals of each major vegetative association and each habitat type. We visited 29 localities (see Table 1) and collected 843 specimens of mammals.

Small rodents were collected using various types of traps including Sherman live-traps, museum specials, and Victor rat traps. Bats were obtained by mist-netting, and by inspecting daytime and nocturnal roosts. Carnivores were taken by shooting or trapping, and lagomorphs and squirrels were obtained by shooting. Some of the larger mammals were difficult to catch and specimens were purchased from residents in the area. The specimens and extensive field notes are deposited in the Texas Cooperative Wildlife Collections of Texas A&M University. Specimens listed as examined are on deposit in that collection unless noted otherwise. Alvarez (1963) summarized information known to that date about mammals in Tamaulipas, and, in general, we have cited in species accounts only pertinent literature published subsequent to his work.

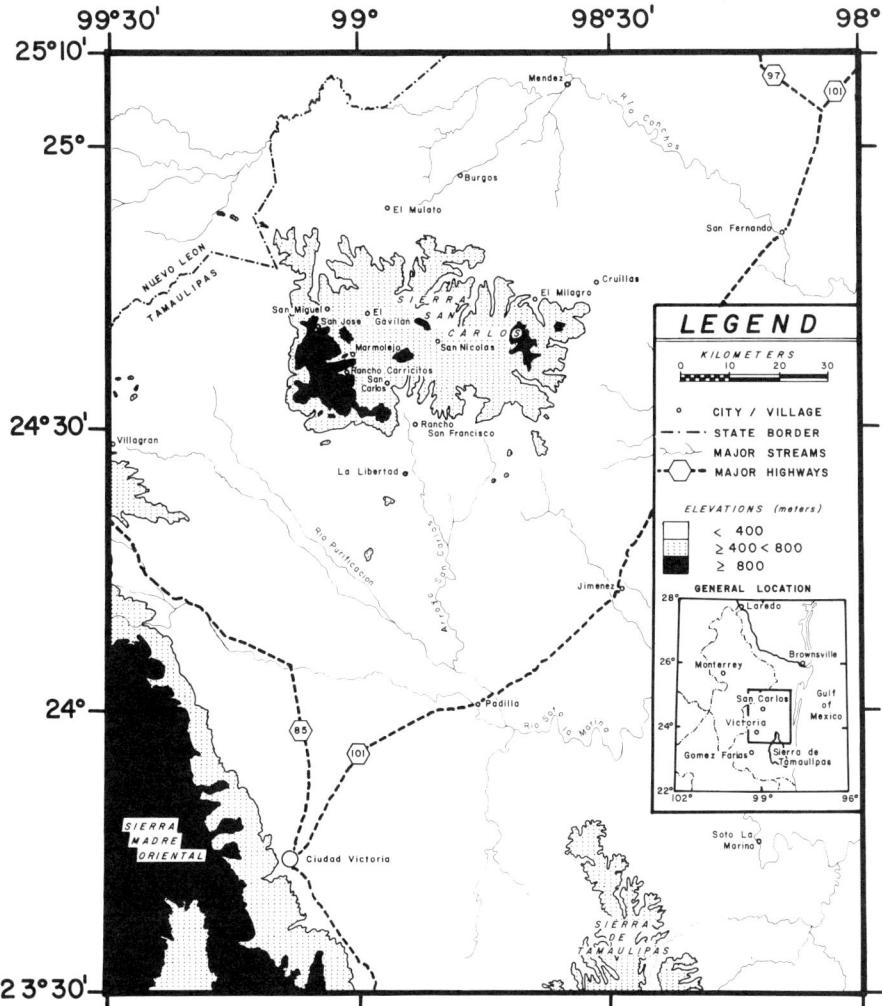

Fig. 1.—The San Carlos Mountains region of northeastern México, indicating major topographic features and important localities.

We also examined specimens, field notes, and photographs by Dice and others of the staff of the Museum of Zoology, University of Michigan (UMMZ); specimens deposited in the Museum of Natural History, University of Kansas (KU); and specimens at the Museum of Natural History, University of Illinois (UIMNH). Skulls were measured with dial calipers; external measurements were recorded by the field collectors. All measurements are in millimeters.

MAJOR HABITAT TYPES OF THE SAN CARLOS REGION

Elevational extremes and attendant climatic contrasts in the San Carlos Mountains have produced a rather wide range of environmental conditions. We recognize the following five major animal habitats:

Xeric Thorn Scrub (Fig. 2, 3).—Characteristic of the driest parts of the region surrounding the mountains is a low, scrub formation with cacti frequently present. The vegetation may be scattered or dense, averaging from one to four meters in height. Xeric thorn scurb is rarely found at elevations exceeding 500 meters and tends to be confined to flat or gently sloping areas, although it occasionally extends onto southerly facing slopes.

Thorn Woodland (Fig. 3, 4).—This formation usually is found at elevations between 350 and 600 meters. It consists of grass-covered, gently sloping or rolling foothill areas in which density of thorn trees varies from sparse to thick enough to form a loose canopy. Thorn trees average from four to 10 meters in height and live oak trees are commonly scattered throughout the area. This formation tends to be broader and extends to slightly higher elevations on north facing slopes. It has been cleared and eliminated in many places for farming.

Deciduous Thickets (Fig. 3, 5).—This formation is found on slopes that normally have a closed canopy and thick leaf litter. Ground cover usually consists of sparse grass and a moderate cover of small herbaceous plants. Thickets may or may not have a thick brierlike understory and are found at elevations between 500 and 900 meters. They are of two types: 1) dry deciduous thickets where the woody vegetation typically is less than 0.2 meters in diameter at the base and the soils are dry, powdery, and strewn with limestone rocks; and 2) moist deciduous thickets where the woody vegetation typically is composed of larger trees (up to 0.4 meters in diameter at the base) and the soils are moist, with humus supporting ferns, mats of algae, and mushrooms on decaying vegetation.

Pine-Oak Forest (Fig. 6).—Pine and oak forest interspersed with grass-lands are characteristic of the Sierra above 800 meters. The oak forest, which is found between 800 and 1400 meters, is typified by oak trees that are 10 to 20 meters tall and often exceed one meter in diameter. This formation is most distinct on south-facing slopes where the canopy is generally open, with a moderately dense understory of tangled vines and shrubs. Pine forest is encountered at elevations above 1000 meters. It is found on upper slopes and crests of the mountains, but occasionally occurs in mesic canyons at lower elevations. Pine forest has little understory or ground cover except for a heavy mat of needles. The transition from oak to pine forest often is gradual and the two formations blend together at many places. For this reason, we have combined them into a single habitat type.

Riparian Forest (Fig. 7).—This formation is found at the edge of streams, reservoirs or other water sources, and includes the associated flood plain. It extends through all of the major habitat types from mountain top to desert.

FIG. 2.—Photograph near Tamaulipeca showing typical xeric thorn scrub habitat. (Courtesy of The Herbarium, The University of Michigan.)

FIG. 3.—Photograph of valley floor at Tamaulipeca showing xeric thorn scrub habitat in foreground grading into thorn woodland in the center and deciduous thickets on the limestone hillsides. (Courtesy of the Herbarium, The University of Michigan.)

FIG. 4.—Photograph near Marmolejo of typical thorn woodland habitat.

FIG. 5.—Photogarph near Tamaulipeca showing typical dry deciduous thickets habitat along the hillsides. (Courtesy of the Herbarium, The University of Michigan.)

Fig. 6.—Photograph of San Carlos Mountains showing pine-oak forest along ridges and upper slopes of mountains. (Courtesy of The Herbarium, The University of Michigan.)

Fig. 7.—Photograph of Arroyo Marmolejo (near Marmolejo) showing typical riparian habitat.

The dominant trees include pecan, sycamore, ebony, and live-oak. Riparian forest frequently has a rather dense understory of briers, tall grasses, and tall weeds. The ground cover typically is humus and frequently is strewn with boulders.

SPECIES ACCOUNTS

Didelphis virginiana californica Bennett

Virginia Opossum (Tlacuache)

Specimens examined (10).—San Carlos Mtns., 1; 0.3-0.5 mi. SW Rancho Carricitos, 4; Mulato, 5 (UMMZ).

Dice (1937:249) reported this opossum as widely distributed in the mountains and plains of the San Carlos region, but we found it only in riparian habitat and suspect that it is most common along water courses. A single specimen was shot in a tree along Arroyo Ebanal near our camp at Rancho Carricitos on 31 December, and four others were brought to that camp by natives. One male, obtained on 12 January 1975, had testis that measured 18 millimeters in length.

We follow Gardner (1973) in assigning our specimens to the subspecies *D. v. californica*. External and cranial measurements of two males and two females, respectively, from Rancho Carricitos are: total length, 802, 820, 732, 749; length of tail, 350, 363, 354, 337; length of hind foot, 62, 64, 48, 52; length of ear, 56, 55, 49, 45; condylobasal length, 112.0, 105.9, 91.4, 91.4; zygomatic breadth, 57.8, 57.4, 44.9, 44.9; length of rostrum, 47.2, 43.2, 36.8, 36.9; length of nasals, 53.4, 45.8, 38.8, 38.6; postorbital constriction, 10.5, 11.4, 9.9, 9.8; and length of maxillary toothrow, 46.5, 44.0, 39.9, 39.8.

Cryptotis parva berlandieri (Baird)

Least Shrew (Musaraña)

Specimen examined (1).—0.3 mi. SW Rancho Carricitos, 1900 ft., 1.

Our specimen was obtained in riparian habitat on 12 January 1976 in a pitfall trap placed near a rock wall. We follow Choate (1970) for use of this trinomial combination. External and cranial measurements of our specimen are: total length, 84; length of tail, 21; length of hind foot, 11; length of ear, 4; greatest length of skull, 17.6; least interorbital constriction, 3.6; mastoid breadth, 8.2; width across M2, 5.1; width across canines, 2.3; depth of skull, 5.1.

Notiosorex crawfordi crawfordi (Coues)

Desert Shrew (Musaraña)

Specimen examined (1).—0.3 mi. SW Rancho Carricitos, 1900 ft., 1.

The specimen is a male collected on 8 January 1976 in riparian habitat. This individual, the third recorded specimen from Tamaulipas, was obtained in the same pitfall trap with the least shrew listed above. The other specimens of *Notiosorex* are from Juamave and Palmillas in southwestern Tamaulipas (Alvarez, 1963:397). One of those was taken in a pine-oak forest, and this shrew probably occupies that habitat in the San Carlos Mountains as well.

We have assigned our specimen to *N. c. crawfordi*, which is the name currently applied to desert shrews in Texas and northern México (Hall, 1981:65). External and cranial measurements of our specimen are: total length, 92, length of tail, 28; length of hind foot, 11; length of ear, 8; greatest length of skull, 18.1; least interorbital constriction, 3.8; mastoid breadth, 8.6; width across M2, 5.2; width across canines, 2.5; depth of skull, 5.0.

Mormoops megalophylla megalophylla Peters
Ghost-faced Bat (Murciélago)

Specimens examined (27).—0.3-0.5 mi. SW Rancho Carricitos, 2; 0.3 mi. SW Rancho Carricitos, 1; 0.5 mi. SW Rancho Carricitos, 1; 0.6 mi. W San Carlos, 2; 6 mi. NE San Miguel, 21.

The ghost-faced bat is distributed throughout Tamaulipas (Alvarez, 1963:399). In the San Carlos Mountains, we netted it over pools of water in riparian habitats along Arroyo Ebanal (January and March) and in xeric thorn scrub habitat near San Carlos (January). Also, several individuals were obtained in a cave near San Miguel on 14-15 January. None was obtained in May, and it is not known whether this species is a spring and summer resident of the mountains.

Mormoops m. megalophylla is the trinominal applied to populations of this species in North America (Smith, 1972). Average external and cranial measurements of eight specimens (five males, three females) from the San Carlos Mountains are (extremes in parentheses): total length, 89.6 (83-95); length of tail, 23.8 (21-28); length of hind foot, 10.5 (9-13); length of ear, 13.2 (11-15); length of forearm, 55.4 (51.8-56.6); greatest length of skull, 15.3 (14.6-15.6); zygomatic breadth, 9.6 (9.4-9.8); postorbital constriction, 5.0 (4.9-5.2); depth of skull, 13.2 (12.9-13.4); length of maxillary toothrow, 7.8 (7.6-7.9); postpalatal length, 6.3 (6.0-6.6).

Leptonycteris nivalis (Saussure)
Mexican Long-nosed Bat (Murciélago)

Specimen examined (1).—0.3-0.5 mi. SW Rancho Carricitos, 1.

This bat is a spring and summer resident of the San Carlos Mountains where it probably roosts in caves and mines. The only record we have is of a male netted over a shallow pool of water on 18 May 1977 in riparian habitat along Arroyo Ebanal; no specimen was taken in January or March. This is the second locality at which this species has been taken in Tamaulipas; the other is a cave near Jimenez (Alvarez, 1963:401), about 40 mi. SE San Carlos.

This species is monotypic (Davis and Carter, 1962). External and cranial measurements of our specimen are: total length, 90; length of forearm, 54.8; greatest length of skull, 26.8; zygomatic breadth, 11.0; postorbital constriction, 4.6; depth of skull, 9.9; length of maxillary toothrow, 8.8; and postpalatal length, 8.4.

Choeronycteris mexicana Tschudi
Long-tongued Bat (Murciélago)

Specimens examined (3).—0.3-0.5 mi. SW Rancho Carricitos, 1; 0.6-1.0 mi. WSW Rancho Carricitos, 2.

Three females of this species were netted on 17 May and 20 May over pools of water in riparian habitat along Arroyo Ebanal. No specimen was taken in January or March, and this species may be resident in the San Carlos Mountains only in spring and summer. One female was pregnant with a single embryo (35 mm. in crown-rump length); another gave birth to a young bat (38 mm.) shortly after capture. The only other records of this bat from Tamaulipas are from the southwestern part of the state along the eastern slopes of the Sierra Madre (Alvarez, 1963:399).

This species is monotypic (Hall, 1981:129). External and cranial measurements of our three specimens are: total length, 105, 85, 80; length of tail, 10, 7, 6; length of ear, 13, 13, 15; greatest length of skull, 29.8, 29.7, 29.8; zygomatic breadth, 10.0, 10.2, 10.1; postorbital constriction, 3.8, 3.6, 3.6; depth of skull, 9.1, 8.9, 9.6; length of maxillary toothrow, 11.4, 11.2, 11.0; postpalatal length, 7.5, 7.2, 8.2.

Desmodus rotundus murinus Wagner
Vampire Bat (Vampiro)

Specimens examined (3).—0.3-0.5 mi. SW Rancho Carricitos, 1; 0.3 mi. SW Rancho Carricitos, 1; 4.9 mi. SSE San Carlos, 1.

Two males, one obtained on 7 January and another on 18 May (testis=7 mm.), were captured in mists nets over pools of water in riparian habitat along Arroyo Ebanal. A female was netted on 5 January in a banana-avocado grove at La Encantado Spring. These bats provide the northernmost occurrence of this species in Tamaulipas (see Alvarez, 1963:405-406) and are the nearest record of this vampire bat to the United States.

We follow Hall (1981:174) in use of the subspecific name *D. r. murinus* for vampire bats from México. External and cranial measurements of our three specimens (female listed first) are: total length, 85, 85, 80; length of hind foot, 18, 18, 18; length of ear, 16, 18, 16; length of forearm, 61.2, 55.4, 58.4; greatest length of skull, 25.2, 24.8, 24.4; zygomatic breadth, 12.8, 12.5, 11.9; postorbital constriction, 5.2, 5.4, 5.4; depth of skull, 13.4, 13.7, 13.2;

length of maxillary toothrow, 3.6, 3.6, 3.4; and postpalatal length, 9.4, 9.4, 9.3.

Myotis velifer incautus (J. A. Allen)

Cave Myotis (Murciélago)

Specimens examined (59).—1.1 mi. WSW Rancho Carricitos, 1; 0.3-0.5 mi. SW Rancho Carricitos, 1; 2.2 mi. SE San Carlos, 4; Cave Huahuiran, 12 mi. N San José, 53 (UMMZ).

We obtained two females on 17-18 March in mist nets over pools of water in riparian habitats along Arroyo Ebanal, and four females on 19-20 March in nets over water in xeric thorn scrub habitat in Arroyo San Carlos. No specimen was obtained in these same places in January and May, but Dice (1937:249) took 54 bats of this species in July from a cave near Huahuiran, which is approximately 12 mi. N San José.

Hayward (1970) reviewed geographic variation in *M. velifer* and assigned specimens from northeastern México to the subspecies *incautus*. Average external and cranial measurements (extremes in parentheses) for six females are: total length, 101.3 (97-106); length of tail, 42.5 (40-46); length of hind foot, 9 (8-10); length of ear, 15.5 (15-16); length of forearm, 43.4 (41.9-44.3); greatest length of skull, 16.5 (16.1-16.8); zygomatic breadth, 10.5 (10.2-11.1); postorbital constriction, 3.9 (3.8-4.0); depth of skull, 7.2 (6.8-7.5); length of maxillary toothrow, 6.3 (6.2-6.4); postpalatal length, 5.0 (4.8-5.3).

Myotis auriculus auriculus Baker and Stains

Southwestern Myotis (Murciélago)

Specimen examined (1).—0.3-0.5 mi. SW Rancho Carricitos, 1.

A single female was collected on 16 May 1977 in a mist net over a pool of water along Arroyo Ebanal. The only other localities where this bat has been taken in Tamaulipas are the type locality in the Sierra de Tamaulipas and Rancho Guayabos (Hall, 1981:206).

We follow Genoways and Jones (1969) for use of this trinomial combination. External and cranial measurements of our specimen are: total length, 90; tail length, 40; hind foot length, 12; ear length, 16; length of forearm, 35.4; greatest length of skull, 15.5; zygomatic breadth, 9.4; postorbital constriction, 3.9; depth of skull, 7.0; length of maxillary toothrow, 6.1; postpalatal length, 4.4. This female is much smaller in cranial measurements (greatest length of skull and length of maxillary toothrow) than topotypes of *M. a. auriculus* as reported by Genoways and Jones.

Myotis californicus californicus (Audubon and Bachman)

California Myotis (Murciélago)

Specimens examined (30).—1.1 mi. WSW Rancho Carricitos, 23; 0.3-0.5 mi. SW Rancho Carricitos, 3; 0.3 mi. SW Rancho Carricitos, 1; 2.8 mi. WSW Rancho Carricitos, 1; 2.2 mi. SE San Carlos, 1; San José, 1 (UMMZ).

TABLE 2.—*External and cranial measurements of the two color phenotypes of* Myotis californicus *occuring in the San Carlos Mountains. Measurements recorded as described by Bogan (1975).*

TCWC no.	Sex	Lenth of forearm	Condylo-premaxillary length	Condylo-canine length	Maxillary toothrow length	Cranial breadth	Cranial depth	Interorbital breadth	Rostral breadth	Rostral length
"Pale Specimens" (*N*=5)										
28789	M	32.4	13.0	12.2	5.2	6.2	5.0	3.5	5.2	5.9
28790	M	33.1		12.2	5.2	6.2	4.5	3.1	4.9	
28771	F	33.4	12.4	11.6	5.2	5.8	4.7	3.0	4.7	5.7
28788	M	31.2	12.6	11.8	5.0	6.0	4.8	3.0	4.8	5.8
28772	F	32.0	12.6	12.2	5.1	6.2	5.0	3.2	4.8	6.0
Mean		32.5	12.6	12.0	5.1	6.1	4.8	3.2	4.9	5.8
"Dark Specimens" (*N*=5)										
28782	F	34.4	12.8	11.9	5.2	6.1	5.1	3.2	5.1	6.0
28783	F	34.8	13.2	12.3	5.3	6.4	5.0	3.4	5.2	6.0
28784	F	35.1	12.9	12.1	5.1	6.2	4.8	3.2	4.9	5.9
28785	F	33.8	12.9	12.1	5.3	6.1	5.0	3.2	5.2	6.0
28787	M	33.0	12.8	11.8	5.1	6.2	5.1	3.2	5.2	5.8
Mean		34.2	12.9	12.1	5.2	6.2	5.0	3.2	5.1	5.9

This bat was common in riparian habitats near Rancho Carricitos. Specimens were taken in January, March, and May in mist nets strung along Arroyo Ebanal and positioned in flyways in the riparian forest. One individual was obtained in pine-oak forest in January and another in xeric thorn scrub habitat along Arroyo San Carlos in March. Dice (1937:249) took a specimen in July from the ruins of an old house at San José. None of our specimens evidenced reproductive activity, but four bats obtained from 2 to 8 January were subadults.

We tentatively follow Bogan (1975) in use of the name combination *M. c. californicus* for bats from the San Carlos region, although there appear to be two distinct kinds in our sample. Bogan (1975), in describing geographic variation in this subspecies, noted that specimens from higher elevations, referred to as "montane" bats, averaged larger in most dimensions and were darker in color than those of the same subspecies from lowland, arid localities. Our sample includes individuals assignable to both of these groups. Five bats are pale in coloration and average small in external and cranial measurements (Table 2). These specimens agree remarkably well with individuals of *M. c. californicus* from Trans-Pecos Texas (Guadalupe and Chisos mountains). The dark-colored bats average larger in most measurements and more closely resemble specimens of *M. c. mexicanus* from southern México (from Tlaxcala, Querétaro, and Chiapas). The systematic implications of these circumstances are difficult to interpret. They could be as described by Bogan or they could result from a winter migration of pale representatives of *M. c. californicus* into the San Carlos Mountains where resident individuals of the dark colored phenotype live throughout the year. Alternatively, the sympatric occurrence of the two forms could be indicative of genetic isolation between them. None of these hypotheses can be supported or refuted with the evidence at hand.

Lasionycteris noctivagans (Le Conte)

Silver-haired Bat (Murciélago)

Specimen examined (1).—2.8 mi. WSW Rancho Carricitós, 1.

One specimen of this bat was obtained, which represents the first record of this species from México (Yates *et al.*, 1976). A nonpregnant female, it was netted on 8 January 1975 over a small pool of water situated at the base of a dry arroyo in pine-oak forest (elevation 1372 m.).

This species is monotypic (Hall, 1981:209). External and cranial measurements of our specimen are: total length, 106; length of tail, 40; length of hind foot, 9; length of ear, 16; length of forearm, 41.2; greatest length of skull, 15.8; zygomatic breadth, 9.8; postorbital constriction, 4.0; depth of skull, 6.6; length of maxillary toothrow, 5.6; postpalatal length, 5.6.

Pipistrellus hesperus maximus Hatfield

Western Pipistrelle (Murciélago)

Specimens examined (9).—2.2 mi. W Rancho Carricitos, 2; 0.3-0.5 mi. SW Rancho Carricitos, 1; 1.1 mi. WSW Rancho Carricitos, 2; 2.2 mi. SE San Carlos, 3; 2.8 mi. WSW Rancho Carricitos, 1.

This is apparently one of the most widespread bats in the San Carlos area inasmuch as we obtained specimens in pine-oak, riparian, and xeric thorn scrub habitats. Specimens were netted in January and March, but not in May. The only other places in Tamaulipas where this species has been recorded are in the extreme southwestern part of the state near Joya Verde (Alvarez, 1963:410) and La Joya Salos (Hall, 1981:212).

We follow Findley and Traut (1970) in assigning our specimens to the subspecies *P. h. maximus*, which occurs widely east of the Continental Divide in the United States and México. Our specimens are noticeably darker in coloration than representatives of this subspecies from Coahuila and Nuevo Leon. Average external and cranial measurements (extremes in parentheses) of five males and two females from the San Carlos Mountains are (males listed first): total length, 71.8 (70-76), 79.5 (78-81); length of tail, 28 (22-32), 33 (32-34); length of hind foot, 5.5 (5-7), 6.0 (6-6); length of ear, 11.8 (11-12), 12.5 (12-13); greatest length of skull, 12.4 (12.2-12.6), 12.8 (12.6-13.0); zygomatic breadth, 7.8 (7.4-8.1), 8.1 (8.0-8.2); postorbital constriction, 3.3 (3.2-3.6), 3.4 (3.2-3.6); depth of skull, 5.6 (5.4-5.8), 5.7 (5.6-5.8); length of maxillary toothrow, 4.3 (4.2-4.5), 4.5 (4.4-4.5); postpalatal length, 4.5 (4.4-4.6), 4.8 (4.6-4.9).

Eptesicus fuscus miradorensis (H. Allen)

Big Brown Bat (Murciélago)

Specimen examined (1).—0.3-0.5 mi. SW Rancho Carricitos, 1.

A single female of this species was captured on 17 March along Arroyo Ebanal, which is considerably north of other localities where this species has been taken in Tamaulipas (Alvarez, 1963:411).

Judging from Hall's (1981:216) distribution map, two subspecies, *E. f. fuscus* and *E. f. miradorensis*, possibly could occur in the San Carlos Mountains. Comparison of our specimen with these two subspecies reveals that it resembles *miradorensis* to a greater degree than *fuscus*, and it accordingly is assigned to the former. External and cranial measurements of the specimen are: total length, 121; length of tail, 50; length of hind foot, 10; length of ear, 18; greatest length of skull, 19.0; zygomatic breadth, 12.8; postorbital constriction, 3.8; depth of skull, 8.8; length of maxillary toothrow, 7.0; postpalatal length, 6.9.

Lasiurus borealis teliotis (H. Allen)

Red Bat (Murciélago)

Specimens examined (18).—1.1 mi. WSW Rancho Carricitos, 6; 0.3-0.5 mi. SW Rancho Carricitos, 5; 0.3 mi. SW Rancho Carricitos, 2; 0.5 mi. SW Rancho Carricitos, 3; 2.8 mi. WSW Rancho Carricitos, 1; 2.2 mi. SE San Carlos, 1.

This is one of the most common and widespread bats occuring in the San Carlos region. We obtained specimens in mist nets over water and along flightways in riparian, xeric thorn scrub, and pine-oak forest habitats. Specimens were taken in January, March, and May, suggesting that this bat is a year-round resident of the area. A female, obtained on 23 May, was pregnant with three fetuses (17 in crown-rump length). The testes of a male captured on 3 January were nonscrotal.

Judging from Hall's (1981:223) distribution map, two subspecies, *L. b. borealis* and *L. b. teliotis*, possibly could occur in the San Carlos Mountains. In order to allocate our material, we compared mean external and cranial measurements of our sample with those of three samples representing these two subspecies (Table 3). A Duncan's Multiple Range Test, used to evaluate statistically significant differences among the four samples, reveals that in most measurements the San Carlos bats (sample 2) are not significantly different from samples of *L. b. teliotis* from western México (sample 4) and the Sierra de Tamaulipas (sample 3). The latter three samples, however, are significantly different in most measurements from sample 1 (*L. b. borealis* from East Texas). For this reason, we herein refer the San Carlos specimens to the subspecies *teliotis*. In most measurements,

TABLE 3.—*Mean external and cranial measurements for four samples of* Lasiurus borealis *representing two subspecies (sample 1 =* L. borealis *from the Big Thicket area of East Texas; sample 2 =* L. borealis *from the San Carlos Mountains; sample 3 =* L. b. teliotis *from the Sierra de Tamaulipas; sample 4 =* L. b. teliotis *from scattered localities in western Mexico). Vertical lines to the right of each array of means connect maximally nonsignificant subsets (as calculated by Duncan's Multiple Range Mean Test) at the 0.05 level.*

Males					Females			
Sample	N	Mean	Grouping		Sample	N	Mean	Grouping
				Total Length				
2	9	101.44			2	9	110.88	
3	6	101.33			1	10	110.60	
1	5	99.60			4	8	108.75	
4	9	95.44			3	10	106.80	
				Tail Length				
3	6	50.50			3	10	54.70	
2	9	47.00			2	9	51.78	
1	5	44.60			4	8	48.25	
4	9	42.44			1	10	45.70	
				Greatest Length of Skull				
1	5	12.97			1	10	13.72	
2	9	12.40			2	9	12.96	
4	9	12.29			3	10	12.82	
3	6	12.23			4	8	12.64	
				Cranial Breadth				
1	5	7.68			1	10	8.08	
3	6	7.44			2	9	7.77	
2	9	7.44			3	10	7.64	
4	9	7.40			4	8	7.63	
				Zygomatic Breadth				
1	5	9.34			1	10	9.98	
3	6	8.85			2	9	9.41	
2	9	8.80			3	10	9.35	
4	8	8.78			4	8	9.11	
				Rostral Breadth				
1	5	5.75			1	10	6.35	
2	9	5.49			3	10	5.87	
3	6	5.49			2	9	5.97	
4	9	5.40			4	8	5.56	
				Length Maxillary Toothrow				
1	5	4.41			1	10	4.78	
3	6	4.05			3	10	4.32	
4	9	4.02			2	9	4.23	
2	9	4.00			4	8	4.18	
				Breadth Across Molors				
1	5	5.93			1	10	6.46	
4	9	5.66			3	10	5.94	
3	6	5.55			2	9	5.93	
2	9	5.52			4	8	5.90	

specimens from the Sierra San Carlos and Sierra de Tamaulipas are intermediate in size between *L. b. borealis* from East Texas (sample 1) and *L. b. teliotis* from western México (sample 4), which would suggest these two subspecies may intergrade in that region of northeastern México.

Lasiurus cinereus cinereus (Palisot de Beauvois)
Hoary Bat (Murciélago)

Specimens examined (15).—1.1 mi. WSW Rancho Carricitos, 11; 0.3-0.5 mi. SW Rancho Carricitos, 2; 0.3 mi. SW Rancho Carricitos, 1; 0.5 mi. SW Rancho Carricitos, 1.

The hoary bat is evidently a common inhabitant of the wooded areas of the San Carlos Mountains during the winter and spring months, where it roosts in trees. We obtained specimens in January, March, and May. Hoary bats were netted over pools of water in riparian habitats along Arroyo Ebanal and in xeric thorn scrub habitat along Arroyo San Carlos. This species is migratory, and one or both sexes may be absent from the area during the warmer months of the year. Adult males were taken in January, March, and May, but adult females were taken only in March. The testes of a male captured on 3 January were nonscrotal, but those of a specimen taken on 19 May measured 5 millimeters in length. We banded and released nine hoary bats on the evenings of 17 to 19 March 1975 and 3 January 1976, but none was recaptured.

The subspecies *cinereus* has a widespread distribution in North America (Hall, 1981:226). Average external and cranial measurements (extremes in parentheses) of 14 adult specimens from the San Carlos Mountains are: total length, 131 (123-138); length of tail, 54.1 (47-61); length of hind foot, 10.2 (8-13); length of ear, 16.8 (13-19); length of forearm, 52.4 (49.7-56.0); greatest length of skull, 16.9 (16.1-18.0); zygomatic breadth, 12.3 (11.9-13.0); postorbital constriction, 5.3 (5.0-5.6); depth of skull, 9.8 (9.5-10.5); length of maxillary toothrow, 5.9 (5.7-6.4); postpalatal length, 7.4 (6.9-7.8).

Lasiurus intermedius intermedius H. Allen
Northern Yellow Bat (Murciélago)

Specimens examined (7).—1.1 mi. WSW Rancho Carricitos, 4; 0.3-0.5 mi. SW Rancho Carricitos, 2; 0.5 mi. SW Rancho Carricitos, 1.

This species apparently is the rarest member of the genus *Lasiurus* that occurs in the San Carlos Mountains. Males and females were netted in January and March over pools of water in riparian habitat along Arroyo Ebanal. No specimens were taken in May. The only other localities where this species has been reported in Tamaulipas are Matamoros and the Sierra de Tamaulipas (Alvarez, 1963:412).

The subspecies *L. i. intermedius*, originally described from Matamoros, Tamaulipas, is widespread from northern Central America and southern México northward along the Gulf Coast to southern Texas and clearly includes material from the San Carlos Mountains (Hall and Jones, 1961). Average external and cranial measurements (extremes in parentheses) of four males and two females are (males listed first): total length, 139 (136-143), 137 (133-141); length of tail, 59.8 (57-62), 57.5 (57-58); length of hind foot, 11.0 (10-12), 11.0 (10-12); length of ear, 16 (13-18), 16 (14-18); length of forearm, 52.4 (48.5-54.8), 49.8 (49.6-50.0); greatest length of skull, 18.4 (17.9-19.1), 18.5 (18.1-19.0); zygomatic breadth, 13.4 (12.8-13.8), 13.8 (13.7-13.9); postorbital constriction, 5.2 (5.1-5.4), 5.2 (4.9-5.6); depth of skull, 9.8 (9.2-10.2), 10.1 (10.0-10.2); length of maxillary toothrow, 6.5 (6.4-6.8), 6.6 (6.2-6.9); postpalatal length, 7.8 (7.4-8.0), 7.3 (7.2-7.4).

Lasiurus ega xanthinus (Thomas)

Southern Yellow Bat (Murciélago)

Specimens examined (20).—2.2 mi. W Rancho Carricitos, 2; 1.1 mi. WSW Rancho Carricitos, 11; 0.3-0.5 mi. SW Rancho Carricitos, 5; 0.5 mi. SW Rancho Carricitos, 1.

This is one of the most common tree bats in the San Carlos Mountains. The only other place in Tamaulipas where it has been recorded is in the Sierra de Tamaulipas (Alvarez, 1963:413). All of our specimens were taken in mist nets set over pools of water in riparian and pine-oak habitats, although this species was more common in the former than it was in the latter habitat type. Specimens of both sexes were obtained in January and March, but only a single male was taken in May. The testes of three males taken in January were small, those of one were 4 mm. in length.

L. e. xanthinus is the trinomial applied by Hall and Jones (1961:91) to populations of this species occurring in México. Average external and cranial measurements (extremes in parentheses) for nine males and six females from the San Carlos Mountains are (males listed first): total length, 115.3 (109-124), 121 (113-131); length of tail, 48.3 (42-57), 49.8 (43-58); length of hind foot, 8 (7-9), 9 (7-11); length of ear, 14 (7-17), 14.3 (12-17); length of forearm, 44.7 (42.1-47.2), 46.3 (44.4-47.6); greatest length of skull, 15.2 (14.8-15.6), 15.5 (15.2-15.9); zygomatic breadth, 10.5 (10.0-10.9), 10.9 (10.5-11.2); postorbital constriction, 4.6 (4.2-4.8), 4.5 (4.3-4.8); depth of skull, 8.4 (8.0-8.8), 8.3 (8.2-8.6); length of maxillary toothrow, 5.2 (5.0-5.4), 5.4 (5.2-5.5); postpalatal length, 6.2 (6.1-6.4), 6.4 (6.2-6.6).

Of other species of bats occurring in the San Carlos Mountains, *Lasiurus ega* most closely resembles *Lasiurus intermedius*. However, *ega* is significantly smaller than *intermedius* in all external and cranial measurements. According to Hall and Jones (1961:79), the two species can be distinguished on the basis of total length, which is always more than 119 in *intermedius* and always less than 119 in *ega*. Our sample of 15 *ega* includes six specimens in which the total length exceeds 119, and this particular measurement, therefore, is not a good feature for distinguishing the two species in the San Carlos Mountains.

Nycticeius humeralis humeralis (Rafinesque)

Evening Bat (Murciélago)

Specimens examined (97).—Rancho Carricitos, 1; 1.1 mi. WSW Rancho Carricitos, 30; 0.3-0.5 mi. SW Rancho Carricitos, 25; 2.8 mi. WSW Rancho Carricitos, 3; 0.5 mi. SW Rancho Carricitos, 38.

This was the most common bat netted in riparian habitats along Arroyo Ebanal. A total of 292 evening bats, including both males and females, was obtained on the following dates: March 1972, 26; January 1975, 103; March 1975, 83; January 1976, 55; and May 1977, 25. Single specimens were captured in the thorn woodland and pine-oak forest habitats, but none was obtained in the xeric thorn scrub. In addition to the individuals listed as examined, we banded 164 evening bats that were netted along Arroyo Ebanal. Sixteen of those subsequently were recaptured. Most had been banded only one or a few nights prior to their recapture, but two bats were taken 10 months after initial capture, and two, 12 months later.

Average testis lengths for males by month were as follows (range in parentheses followed by sample size): January, 6 (4-9) 5; March 5.8 (5-7) 5; May, 5.2 (4-6) 5. None of the females captured in January or March evidenced reproductive activity, but five adult females captured in early May were pregnant. Four of these contained two fetuses each that measured 10, 12, 13, and 14 in crown-rump length; one contained a single fetus that measured 13 mm.

Two subspecies of *Nycticeius humeralis* are known from the southern part of its range. Specimens from Texas and Matamoros, Tamaulipas, have been assigned to *N. h. humeralis*, whereas specimens from southern Tamaulipas, Nuevo Leon, and San Luis Potosí are assigned to *N. h. mexicanus* (Hall, 1981:227). To assess the subspecific affinities of the San Carlos material, we compared a sample of evening bats from the mountains with a sample of *N. h. humeralis* from Nacogdoches, Texas, and a sample of *N. h. mexicanus* from San Luis Potosí (Table 4). With the exception of forearm length, there is little appreciable difference between the samples from Nacogdoches and San Luis Potosí. Specimens from San Carlos average larger in most measurements than do either of the other samples. Given this pattern of variation, we seriously question the validity of recognizing two subspecies of *N. humeralis* in the southern part of the range of the species and, hence, have referred the San Carlos material to the nominate race *N. h. humeralis*.

Plecotus townsendii australis Handley

Townsend's Big-eared Bat (Murciélago)

Specimens examined (3).—2.2 mi. SE San Carlos, 3.

We obtained two males and one female on 20 March in a mist net strung over a pool of water along Arroyo San Carlos in xeric thorn scrub habitat.

TABLE 4.—*Mean (± two standard errors) of external and cranial measurements for three samples of* Nycticeius humeralis *from Texas and northeastern México.*

Locality (N)	Forearm length	Skull length	Zygomatic breadth	Interorbital constriction	Length maxillary toothrow	Width across molars
			Males			
Nacogdoches Co., Texas (19)	35.20+1.35	14.25+0.40	9.78+0.28	3.90+0.15	5.04+0.22	6.34+0.15
San Carlos Mountains (19)	35.01+0.80	14.57+0.17	10.06+0.15	3.87+0.08	5.27+0.07	6.52+0.07
San Luis Potosí (23)	33.48+0.40	14.34+0.08	9.81+0.11	3.82+0.06	5.00+0.07	6.17+0.07
			Females			
Nacogdoches Co., Texas (36)	36.18+0.98	14.55+0.33	10.01+0.26	3.90+0.12	5.19+0.22	6.43+0.17
San Carlos Mountains (11)	35.01+0.80	14.82+0.22	10.21+0.10	3.96+0.07	5.31+0.08	6.66+0.07
San Luis Potosí (16)	34.16+0.70	14.51+0.15	10.10+0.12	3.91+0.06	5.15+0.09	6.35+0.11

Those specimens represent the second record of occurrence for this species in Tamaulipas, the other being from near Miquihuana in the extreme southwestern part of the state (Hall, 1981:235).

Handley (1959) revised the genus *Plecotus* and assigned specimens from southern Coahuila and San Luis Potosí to *P. t. australis.* Our specimens compare favorably with big-eared bats from those areas and, hence, we assign them to that subspecies. External and cranial measurements of our specimens (males listed first) are: total length, 105, 98, 110; length of tail, 47, 46, 50; length of hind foot, 11, 10, 10; length of ear, 34, 36, 35; length of forearm, 43.4, 42.2, 43.6; greatest length of skull, 16.8, 16.5, 17.2; zygomatic breadth, 8.8, 8.6, ——; postorbital constriction, 3.6, 3.6, 3.6; depth of skull, 6.2, 6.0, 6.2; length of maxillary toothrow, 5.1, 5.0, 5.4; postpalatal length, 6.2, 6.2, 6.3.

Antrozous pallidus pallidus (Le Conte)

Pallid Bat (Murciélago)

Specimens examined (17).—0.3-0.5 mi. SW Rancho Carricitos, 2; 0.3 mi. SW Rancho Carricitos, 2; 2.2 mi. W Rancho Carricitos, 1; 2.8 mi. WSW Rancho Carricitos, 2; 0.5 mi. SW Rancho Carricitos, 3; 1.1 mi. WSW Rancho Carricitos, 7.

Pallid bats were netted in the riparian habitats along Arroyo Ebanal and at higher elevations in pine-oak forest west of Rancho Carricitos. Twenty-eight individuals, representing both sexes, were obtained in the following months: January, 12 (11 males, 1 female); March, 15 (13, 2); and May, 1 (1, 0). The fact that most specimens were males might indicate that females were in nursery colonies elsewhere, at least in spring months. One male taken in January had testes that measured 7 mm. in length, and three males collected in March had scrotal testes. In addition to the individuals listed as examined, we banded 11 along Arroyo Ebanal on the nights of 15 to 18 March 1975 and 7 January 1976. None of the banded bats was recaptured.

Martin and Schmidly (1982) reviewed geographic variation in *Antrozous pallidus* and concluded that populations from Tamaulipas are assignable to the subspecies *pallidus*. Previously, Baker (1967) had referred specimens from northeastern México, including the San Carlos Mountains, to *A. p. obscurus*. Average external and cranial measurements (extremes in parentheses) of 10 specimens from the vicinity of Rancho Carricitos are: total length, 110.5 (101-124); length of tail, 46.2 (40-54); length of hind foot, 11 (8-13); length of ear, 27.7 (26-29); length of forearm, 52.1 (49.8-54.6); greatest length of skull, 20.6 (20.2-21.0); zygomatic breadth, 12.2 (11.8-12.6); postorbital constriction, 4.0 (3.8-4.2); depth of skull, 8.6 (8.4-8.8); length of maxillary toothrow, 7.0 (6.8-7.3); postpalatal length, 6.8 (6.4-7.3).

Tadarida brasiliensis mexicana (Saussure)
Brazilian Free-tailed Bat (Murciélago)

Specimens examined (85).—1.1 mi. WSW Rancho Carricitos, 32; 0.3-0.5 mi. SW Rancho Carricitos, 3; 0.3 mi. SW Rancho Carricitos, 2; 2.2 mi. SE San Carlos, 1; 6 mi. NE San Miguel, 39; 0.5 mi. SW Rancho Carricitos, 8.

Aside from the evening bat, *Nycticeius humeralis*, this was the most common bat netted in riparian habitat along Arroyo Ebanal. It also was taken in xeric thorn scrub habitat near San Carlos, and in the Guano Cave near San Miguel. Specimens were obtained in January, March, and May. Average testis lengths for males by month were as follows (range in parentheses followed by sample size): January, 4.4 (1-6) 5; March, 8.2 (5-16) 4. Several subadults were captured at Rancho Carricitos on 7 and 8 January. None of the females collected was pregnant.

In addition to examining the individuals listed, we banded 148 along Arroyo Ebanal (near Rancho Carricitos) and Arroyo San Carlos (near San Carlos). Only five bats were recaptured, and those were banded one or two nights previously. One of the bats, banded on 18 March 1975, was recaptured twice (19 and 20 March 1975).

Populations of this species in the western United States and most of México have been assigned to *T. b. mexicana* (Hall, 1981:242). We follow this arrangement, although it has been questioned by some recent investigators (Cockrum, 1969). Average external and cranial measurements of 15 specimens (nine females, six males) from near Rancho Carricitos are (extremes in parentheses): total length, 96.1 (90-102); length of tail, 36.2 (31-46); length of hind foot, 8.8 (7-10); length of ear, 17.3 (14-20); length of forearm, 42.9 (41.3-44.5); greatest length of skull, 16.7 (15.9-17.5); zygomatic breadth, 9.9 (9.5-10.3); postorbital constriction, 3.8 (3.5-4.0); depth of skull, 7.4 (7.1-7.8); length of maxillary toothrow, 5.8 (5.6-6.0); postpalatal length, 7.0 (6.8-7.6).

Tadarida macrotis (Gray)

Big Free-tailed Bat (Murciélago)

Specimens examined (6).—0.3-0.5 mi. SW Rancho Carricitos, 1; 1.1 mi. WSW Rancho Carricitos, 5.

Our records constitute the first reports of this species from Tamaulipas (Baumgardner *et al.*, 1977). We netted six specimens, five males and one female, over pools of water in riparian habitat along Arroyo Ebanal on five nights during January. A male obtained on 9 January had testes that measured 5 mm. in length. This species is well known for its wandering tendencies (Barbour and Davis, 1969), and evidently a few, possibly migrant, indivduals occur in the San Carlos Mountains during the winter.

This species is monotypic (Hall, 1981:245). Average external and cranial measurements of our specimens are (extremes in parentheses): total length, 137 (128-153); length of tail, 53.8 (47-62); length of hind foot, 12 (10-13); length of ear, 25.3 (18-32); length of forearm, 59.8 (57.1-62.4); greatest length of skull, 23.6 (23.2-23.9); zygomatic breadth, 12.7 (12.5-12.9); postorbital constriction, 4.0 (3.9-4.2); depth of skull, 9.8 (9.6-10.0); length of maxillary toothrow, 8.5 (8.3-8.8); postpalatal length, 10.2 (9.8-10.6).

Dasypus novemcinctus mexicanus Peters

Nine-banded Armadillo (Armadillo)

Specimens examined (3).—5 mi. N Tamaulipeca, 1 (UMMZ); Marmolejo, 2 (UMMZ).

The armadillo supposedly occurs throughout Tamaulipas (Alvarez, 1963:418), but surprisingly we did not secure any specimens or document any sign of it in the San Carlos Mountains. However, long-term residents told us they are common throughout the area and live in all habitat types. Dice (1937:256) reported armadillos from Tamaulipeca and Marmolejo in thorn woodland, deciduous thickets, and riparian forest habitat.

We follow Hall (1981:284) in assigning armadillos from the San Carlos area to *D. n. mexicanus*.

Sylvilagus floridanus chapmani (J. A. Allen)

Eastern Cottontail (Conejo)

Specimens examined (25).—Rancho Carricitos, 1; 0.3 mi. SW Rancho Carricitos, 2; 0.3-0.5 mi. SW Rancho Carricitos, 8; Tinaja, 4; 0.7 mi. E Tinaja, 1; 12 mi. NW San Carlos, 1300 ft., 3 (KU); Tamaulipeca, 6 (UMMZ).

The eastern cottontail is abundant in the broad valleys extending into the mountains wherever there is sufficient cover to provide daytime hiding places. All of our specimens were obtained in riparian forest near Rancho Carricitos and in thorn woodland habitat adjacent to Tinaja. Cottontails, presumably of this species, also were observed at higher elevations in pine-oak forest. Dice (1937:255) obtained specimens at Tamaulipeca in mesquite, granjeno, and live-oak plant associations, which would be included in our deciduous thicket habitat.

Five males obtained in January had testicular measurements of 34, 14, 40, 34, and 55. None of the females obtained in January was pregnant; at Tamaulipeca, Dice (1937:255) reported a female containing three 15 to 20-millimeter fetuses on 24 July and another with three fetuses of about the same size on 26 July. A female obtained 12 mi. NW San Carlos on 23 August was pregnant with two 15-millimeter fetuses, and a male taken on the same date and at the same place had testes that measured 22 in length.

The subspecies *S. f. chapmani*, to which our specimens are assigned, includes material from southern Texas and northeastern México (Hall, 1981:303). Average external and cranial measurements (extremes in parentheses) of a sample of 15 adults from the San Carlos region are: total length, 372.8 (310-482); length of tail, 40.4 (34-48); length of hind foot, 81.3 (75-88); length of ear, 57.2 (45-62); greatest length of skull, 66.3 (62-70); zygomatic breadth, 33.0 (31.5-34.6); length of nasals, 26.6 (23.4-30.6); interorbital constriction, 11.9 (10.5-13.4); mastoid breadth, 28.2 (27.3-29.2); length of maxillary toothrow, 12.1 (11.0-13.1); depth of skull, 30.0 (28.0-30.1).

Sylvilagus audubonii parvulus (J. A. Allen)
Desert Cottontail (Conejo)

Specimen examined (1).—Mulato, 1 (UMMZ).

The desert cottontail is known in the San Carlos region on the basis of a single specimen collected by Dice (1937:256) at El Mulato in the mesquite association of the xeric thorn scrub vegetation belt. This species reaches the eastern limits of its distribution in western Tamaulipas, and apparently it is not very common in this part of México. We did not obtain any specimens in the foothills or montane habitats of the San Carlos Mountains.

We assigned our specimen to the subspecies *S. f. parvulus*, which was described based upon material from Hidalgo, and currently is thought to occur over much of northeastern México and southern Texas (Hall, 1981:309). External and cranial measurements of the single specimen, an adult male, from El Mulato are as follows: total length, 362; length of tail, 33; length of hind foot, 74; length of ear, 63; basilar length of skull, 57.8; zygomatic breadth, 32.7; breadth of braincase, 25.8; length of nasals, 23.0; depth of rostrum, 12.8; length of maxillary toothrow, 12.2; length of auditory bullae, 12.6.

Lepus californicus merriami Mearns
Black-tailed Jack Rabbit (Liebre)

Specimens examined (7).—12 mi. NW San Carlos, 1300 ft., 2 (KU); Tamaulipeca, 1 (UMMZ); San Miguel, 2 (UMMZ); Mulato, 2 (UMMZ).

Dice (1937:255) reported jack rabbits to be numerous on the lower plains and to extend into the larger valleys and lower foothills at the northern part of the San Carlos region. He saw individuals at El Gavilán, San Miguel, Cruillas, and El Mulato, and purchased specimens at Tamaulipeca; those places are in the xeric thorn scrub and thorn woodland habitat. We obtained no specimens during our work in the mountains, nor did we observe jack rabbits in the foothills and mountainous area surrounding the interior valley. Futhermore, long-term residents of the area told us that they do not occur at Marmolejo and Gavilán in the valley. We saw individuals in the xeric thorn scrub habitat in the vicinity of La Libertad and Rancho San Francisco in the southern part of the area.

We have assigned our specimens to the taxon *Lepus californicus merriami*, which is the name used by Hall (1951*b*:185) for the specimen from Tamaulipeca. External and cranial measurements of three adult females and a male, respectively, from the San Carlos area are as follows: total length, 588, 573, ——, ——; length of tail, 76, 61, ——, ——; length of hind foot, 123, 117, ——, ——; length of ear, 117, 113, ——, ——; basilar length of skull, 87.4, 86.5, 88.1, 84.4; zygomatic breadth, 43.1, 43.4, 43.2, 44.8; breadth of braincase, 31.5, 32.6, 30.6, 32.8; length of nasals, 37.8, 36.6, 35.7, 33.0; depth of rostrum, 21.6, 21.1, 22.7, 20.8; length of maxillary toothrow, 17.4, 16.4, 15.8, 16.5; length of auditory bullae, 14.1, 13.6, 13.8, 12.8.

Spermophilus mexicanus parvidens Mearns

Mexican Ground Squirrel (Tuza)

Specimens examined (15).—Tamaulipeca, 11 (UMMZ); Mulato, 4 (UMMZ).

We did not take any specimens of the Mexican ground squirrel in our work, but Dice (1937: 251) reported specimens from the following localities and plant associations in xeric thorn scrub and thorn woodland habitats: El Mulato, mesquite association; Tamaulipeca, granjeno association; and El Gavilán, granjeno association. One of us (Hendricks) saw a ground squirrel of this species at the edge of a pasture along the road between Marmolejo and Gavilán. Residents of the area told us these ground squirrels are common in the interior valley and around Rancho San Francisco where they live along the edges of the fields.

We follow Hall (1981:394) in assigning our specimens to the subspecies *S. m. parvidens* on geographical grounds. Cranial measurements of four adults (three males and a female, respectively) from Tamaulipeca are as follows: greatest length of skull, 44.4, 44.2, 45.0, 43.8; zygomatic breadth, 26.2, 26.3, 25.5, 25.2; length of rostrum, 16.8, 16.8, 17.6, 16.0; interorbital constriction, 9.6, 9.2, 8.9, 9.2; mastoid breadth, 20.4, 19.8, 19.5, 19.6; length of nasals, 14.2, 15.2, 15.8, 13.5; length of maxillary toothrow, 7.6, 7.9, 8.4, 8.5; depth of skull, 17.0, 16.8, 16.4, 17.2.

Sciurus aureogaster aureogaster Cuvier

Red-bellied Squirrel (Ardilla)

Specimens examined (18).—0.3-0.5 mi. SW Rancho Carricitos, 1; Rancho Carricitos, 1; 0.3 mi. SW Rancho Carricitos, 11; 2.2 mi. W Rancho Carricitos, 3; 0.5 mi. SW Rancho Carricitos, 1; Marmolejo, 1 (UMMZ).

We were successful in shooting several red-bellied squirrels in riparian habitats near Rancho Carricitos and a few at higher elevations in pine-oak forest. We purchased several specimens from residents living near Rancho Carricitos, and Dice (1937:252) reported purchasing a melanistic female at Marmolejo. Only three of our specimens exhibited the melanistic pelage phase; the remainder were of the frosted-gray phase.

Two females, taken on 5 and 7 January, evinced no gross reproductive activity. Two males had testicular lengths of 30 and 8 on 1 and 6 January.

We have followed the arrangement of Musser (1968), who recognized only two subspecies of this squirrel, with the name *S. a. aureogaster* being applied to populations from northeastern México. Average external and cranial measurements (extremes given in parentheses) of 16 individuals from the San Carlos region are: total length, 490.3 (400-524); length of tail, 241.5 (188-269); length of hind foot, 60.2 (53-66); length of ear, 29.1 (21-34); greatest length of skull, 57.2 (55.2-59.9); zygomatic breadth, 32.8 (31.7-34.8); length of nasals, 17.8 (16.6-20.4); interorbital constriction, 18.4 (16.8-19.3); mastoid breadth, 26.6 (25.6-27.8); length of maxillary toothrow, 10.7 (9.6-11.6); depth of skull, 25.1 (24.1-26.2).

Sciurus alleni Nelson

Allen's Squirrel (Ardilla)

Specimens examined (31).—1.5 mi. WSW Rancho Carricitos, 5; Rancho Carricitos, 4; 0.3 mi. SW Rancho Carricitos, 2; 2.2 mi. W Rancho Carricitos, 9; 2.8 mi. WSW Rancho Carricitos, 1; San José, 5 (UMMZ); Marmolejo, 2 (UMMZ); Cruillas, 3 (UMMZ).

There seems to be some ecological displacement between Allen's squirrel and the red-bellied squirrel, *S. aureogaster*. The former is more abundant at higher elevations in pine-oak forest, whereas the latter seems to prefer riparian forests at lower elevations in the mountains. We shot several Allen's squirrels in pine-oak forest above Rancho Carricitos, and natives brought us specimens taken in the same region. Dice (1937:252) either sighted or obtained specimens in forests of madrona, oak, and pine at San José, in a forest of live oak and pecan at Marmolejo, and in the live-oak forest at El Milagro.

Two males, taken on 12 January and 20 March, had testes that were 4 and 19, respectively, in length. Two females taken on 12 January evinced no gross reproductive activity. Dice (1937:252) reported a female containing four 20-millimeter fetuses on 21 July and another with four 37-millimeter fetuses on 10 August; a nest in a hollow branch of a jaboncillo tree contained young on 23 August.

This species is monotypic (Hall, 1981:430). Average external and cranial measurements (extremes in parentheses) of 11 adults from the San Carlos Mountains are: total length, 427.7 (380-459); length of tail, 189.1 (135-212); length of hind foot, 57.1 (51-61); length of ear, 30.2 (22-33); greatest length of skull, 57.5 (56.1-59.4); zygomatic breadth, 32.8 (31.8-34.1); length of nasals, 18.8 (17.9-19.6); interorbital constriction, 17.3 (16.4-18.2); mastoid breadth, 26.6 (25.4-27.4); length of maxillary toothrow, 9.9 (9.5-10.2); depth of skull, 24.5 (23.9-25.2).

Perognathus flavus merriami J. A. Allen

Silky Pocket Mouse (Ratón)

Specimens examined (22).— 12 mi. NW San Carlos, 1300 ft., 1 (KU); Mulato, 20 (UMMZ); 6.2 mi. NW San Carlos, 1 (UIMNH).

The silky pocket mouse has been taken only in xeric thorn scrub and thorn woodland habitats; apparently, it does not occur in deciduous thicket, riparian, or pine-oak forest habitats. Dice (1937:257) reported specimens from xeric thorn scrub and thorn woodland habitats at El Mulato and west of Cruillas. We obtained a specimen in thorn woodland habitat near Marmolejo.

Wilson (1973) has demonstrated that two nominal taxa of silky pocket mice—*P. flavus* and *P. merriami*—are conspecific. Under this arrangement, specimens from Tamaulipas and the San Carlos area should be referred to the taxon *P. f. merriami*. Average external and cranial measurements (extremes in parentheses) of 6 specimens from Mulato are: total length, 119 (112-124); length of tail, 58.2 (54-61); length of hind foot, 15.7 (14-17); length of ear, 6 (5-7); greatest length of skull, 20.8 (19.6-21.4); zygomatic breadth, 11.2 (10.7-11.7); length of rostrum, 8.3 (7.6-8.7); length of nasals, 7.4 (6.6-7.8); least interorbital constriction, 4.4 (4.2-4.6); mastoid breadth, 11.4 (10.7-12.0); length of maxillary toothrow, 3.0 (2.8-3.2); depth of skull, 7.7 (7.4-8.0).

Perognathus hispidus hispidus Baird

Hispid Pocket Mouse (Ratón)

Specimens examined (3).—Mulato, 3 (UMMZ).

This is another species reported by Dice (1937:253) from the lower plains (thorn woodland habitat) surrounding the San Carlos Mountains. He collected three specimens at El Mulato; one was taken along cattails growing in the seepage from a reservoir, one in arid thorn shrubs at the edge of the sedge association below the reservoir, and another in the mesquite association near a dry, gravelly arroyo. We did not obtain this species in the foothills or montane habitats during our work in this area.

We follow Hall (1981:546) in assigning our specimens to the subspecies *P. h. hispidus* on geographic grounds. External and cranial measurements of two adults (male and female) from Mulato are as follows: total length, 175, 183; length of tail, 89, 88; length of hind foot, 22, 24; length of ear, 10, 10; greatest length of skull, 26.1, 27.8; zygomatic breadth, 13.2, 14.6; length of rostrum, 10.5, 11.4; length of nasals, 9.2, 9.1; least interorbital constriction, 6.5, 6.6; mastoid breadth, 12.4, 14.5; length of maxillary toothrow, 4.2, 4.4; depth of skull, 9.7, 10.4.

Liomys irroratus texensis Merriam

Mexican Spiny Pocket Mouse (Ratón)

Specimens examined (54).—0.4 mi. NW Rancho Carricitos, 1; 0.3 mi. W Rancho Carricitos, 4; 0.3-0.5 mi. SW Rancho Carricitos, 18; 0.6-1.0 mi. WSW Rancho Carricitos, 3; 1.1 mi. WSW Rancho Carricitos, 2; 2.6 mi. WNW San Carlos, 1600 ft., 7; 3.4 mi. NNE San Carlos, 1; 4.3 mi. SE San Carlos, 1; 4.9 mi. SSE San Carlos, 1; San José, 1 (UMMZ); Marmolejo, 7 (UMMZ); Mulato, 1 (UMMZ); 11 mi. SW Cruillas, 2 (UMMZ); 6.2 mi. NW San Carlos, 5 (UIMNH).

This species is one of the most common and ubiquitous rodents in the San Carlos region. We trapped 113 spiny pocket mice in the following habitat types: riparian forest, 57; thorn woodland, 43; pine-oak forest, 6; deciduous thickets, 3; and xeric thorn scrub, 1. Dice (1937:252) reported specimens from the following localities and plant associations (our habitats in parentheses): San José, granjeno association (deciduous thickets); Marmolejo, live-oak association (riparian forest); El Milagro, live-oak association (riparian forest); El Mulato, mesquite association (xeric thorn scrub). Each of these places is located in the valleys leading into the mountains.

Trapping records clearly indicate that this species is most common in the riparian and thorn woodland habitats, and in January 1976 we established grids (each encompassing 0.61 hectares) in each of those habitats for the purpose of making population estimates of this species. The grids consisted of Sherman live traps placed in five parallel rows, 13.7 m. apart, with 10 traps per row at a distance between adjacent traps of 9.2 m. The Schnabel formula was used to calculate daily population estimates for each of the two grids. The grid in the riparian forest was operated from 2 to 6 January 1976. Only three *Liomys* were captured, and no recaptures were obtained; therefore, it was impossible to calculate population estimates. The grid in the thorn woodland habitat was operated from 10 to 13 January 1976. Eleven *Liomys* were captured and four of these were recaptured, giving a population estimate of nine individuals per 0.61 hectares.

Testicular lengths of seven males collected in May averaged 21 (range, 14-25), and a single female collected on 9 January was lactating. Dice (1937) captured a pregnant female with three small fetuses on 9 August at Marmolejo.

We follow Genoways (1973) in applying the name *L. i. texensis* to our specimens. Average external and cranial measurements (extremes in parentheses) of a series of 20 adult specimens are: total length, 237.3 (215-275); length of tail, 117.3 (104-135); length of hind foot, 27.7 (21-30); length of ear, 13.9 (10-18); greatest length of skull, 31.2 (29.8-32.3); zygomatic breadth, 14.9 (14.2-15.8); length of nasals, 12.2 (10.8-13.0); interorbital constriction, 7.7 (7.2-8.2); mastoid breadth, 15.0 (14.2-16.2); length of maxillary toothrow, 4.8 (4.4-5.1); depth of skull, 10.6 (10.2-11.0).

Reithrodontomys fulvescens intermedius J. A. Allen
Fulvous Harvest Mouse (Ratón)

Specimens examined (9).—2.2 mi. W Rancho Carricitos, 3; 0.5 mi. SW Rancho Carricitos, 2; 0.6-1.0 mi. WSW Rancho Carricitos, 1; 2.6 mi. WNW San Carlos, 1600 ft., 1; Tamaulipeca, 1 (UMMZ); Mulato, 1 (UMMZ).

The fulvous harvest mouse probably occurs throughout the mountains wherever grass occurs. We obtained our specimens in dense thickets of grass in riparian, pine-oak, and thorn woodland habitats. Dice (1937:253) reported only two specimens of this species, one in high grass growing along the edge of a dry steam bed just below Tamaulipeca (thorn woodland habitat) and the other in a mixed growth of low grass, sedges, and rushes below a reservoir near El Mulato (thorn woodland).

A female collected on 25 May contained no fetuses. Two males, collected on the same day, had testicular lengths of 5 mm.

We follow Hall (1981:646) in assigning our specimens to the subspecies *R. f. intermedius*. This name was originally based on material from Cameron County, Texas, and is currently applied to specimens from much of northeastern México and South Texas. External and cranial mesurements for five of our specimens are: total length, 168, 159, 155, 187, 156; length of tail, 94, 87, 85, 111, 91; length of hind foot, 19, 19, 17, 21, 20; length of ear, 15, 14, 15, 17, 15; greatest length of skull, 21.1, 21.6, 21.0, 22.3, 20.7; zygomatic breadth, 11.1, 11.0, 11.0, 10.9, 11.0; length of rostrum, 7.6, 7.7, 7.4, 8.2, 7.0; length of nasals, 7.8, 8.4, 7.9, 8.4, 7.2; least interorbital constriction, 3.2, 3.4, 3.2, 3.4, 3.4; length of maxillary toothrow, 3.4, 3.7, 3.6, 3.2, 3.3; depth of skull, 8.5, 8.3, 8.8, 8.7, 8.0.

Peromyscus leucopus texanus (Woodhouse)
White-footed Mouse (Ratón)

Specimens examined (66).—0.3 mi. SW Rancho Carricitos, 9; 0.3-0.5 mi. SW Rancho Carricitos, 11; 0.9 mi. WSW Rancho Carricitos, 1; 0.5 mi. SW Rancho Carricitos, 2; 1.1 mi. WSW Rancho Carricitos, 2; 2.6 mi. WNW San Carlos, 1600 ft., 25; 4.9 mi. SSE San Carlos, 4; San José, 3 (UMMZ); Marmolejo, 2 (UMMZ); Mulato, 3 (UMMZ); 6.2 mi. NW San Carlos, 4 (UIMNH).

This species is not nearly so common as are *P. pectoralis* and *P. boylii* in the San Carlos Mountains. We secured specimens in the riparian (22 individuals) and thorn woodland (28) habitats near Rancho Carrcitos, as well as a banana-avocado grove near La Encontada Spring (4). Dice (1937:253) reported this species from El Mulato (mesquite-scrub association and cattail thicket in thorn woodland), San José (granjeno association in thorn woodland) and Marmolejo (live-oak association in riparian forest).

A female taken on 5 January was not pregnant, but one collected on 19 May near Rancho Carricitos had four 15-millimeter fetuses. A female trapped on 8 January at La Encontada Spring was lactating. Dice (1937) obtained a pregnant female on 9 July at San José with two-millimeter fetuses. Testes lengths for males were as follows: January, 5, 5, 6, 7, 10; May, 5, 15.

We follow Hall (1982:687) in assigning our specimens to *P. 1. texanus*. Average external and cranial measurements (extremes in parentheses) of a series of 10 adults from the vicinity of Rancho Carricitos are: total length, 170 (151-192); length of tail, 80.1 (74-90); length of hind foot, 21 (20-24); length of ear, 17.1 (15-19); greatest length of skull, 25.4 (24.2-26.4); zygomatic breadth, 13.2 (12.4-13.8); length of nasals, 9.3 (8.2-10.3); interorbital constriction, 4.1 (3.9-4.3); mastoid breadth, 11.2 (10.8-11.6); length of maxillary toothrow, 3.7 (3.5-4.0); depth of skull, 9.1 (8.6-9.5).

Peromyscus pectoralis collinus Hooper

White-ankled Mouse (Ratón)

Specimens examined (138).—1 mi. WNW Rancho Carricitos, 11; 0.7 mi. W Rancho Carricitos, 9; 0.4 mi. W Rancho Carricitos, 3; 0.3 mi. W Rancho Carricitos, 7; 0.3 mi. SW Rancho Carricitos, 25; 0.3-0.5 mi. SW Rancho Carricitos, 4; 1 mi. WSW Rancho Carricitos, 5; 1.1 mi. WSW Rancho Carricitos, 1; 2.2 mi. W Rancho Carricitos, 1; 3.4 mi. NNE San Carlos, 2; 4.9 mi. SSE San Carlos, 3; San José, 20 (UMMZ); La Vegonia, 2 (UMMZ); Tamaulipeca, 8 (UMMZ); Marmolejo, 15 (UMMZ); 6.2 mi. NW San Carlos, 5 (UIMNH).

This species is common in the foothills of the San Carlos Mountains where it occurs in several habitats with the brush mouse, *Peromyscus boylii*. The distributions of the two species, as reflected by our trapping results in the various habitat types, are depicted in Table 5. Although large numbers of both species were captured in the riparian forest, some degree of ecological separation is indicated. *Peromyscus pectoralis* demonstrates an affinity for deciduous thickets and thorn woodland, whereas *P. boylii* is dominant in oak and pine forests.

Of 29 adult males collected in May, 26 had scrotal testes that averaged 12.7 (range 10-22) in total length. Twelve adult females also were taken in May. Four were pregnant, each with three 13-millimeter fetuses; two others were lactating. No individuals obtained in other months showed evidence of reproductive activity.

P. pectoralis and *P. boylii* are so similar morphologicaly that they are often difficult to distinguish, and they have been suspected of interbreeding by some investigators (Dalquest, 1953; Hooper, 1952). The most distinctive morphological difference between them is in the baculum of males (Clark, 1953). In *pectoralis*, that structure is capped by a long, attenuate cartilaginous spine; *boylii* has only a small rounded cone. This feature, however, is of limited utility because it is present in the male only and often is not preserved in museum specimens. Hooper (1952) and Modi (1978) studied the two species in the San Carlos Mountains and documented several qualitative

TABLE 5.—*Distribution of* P. pectoralis *and* P. boylii *in the habitat types of the San Carlos Mountains.*

Habitat	No. *pectoralis* collected	No. *boylii* collected
Riparian Forest	43	46
Thorn Woodland	10	0
Riparian/		
Thorn Woodland Ecotone	3	3
Deciduous Thickets	26	0
Pine-Oak Forest	3	45
Totals	85	94

and quantitative differences of the external pelage and cranium, some of which are shown in Table 6. As compared to *boylii, pectoralis* generally is larger in external measurements and has a longer skull, with a more inflated braincase and longer interparietal, but with a shorter and narrower rostrum as well as shorter nasal bones and maxillary toothrow. Using a discriminant analysis, Modi (1978) found that all individuals of the two species could be distinguished using conventional external and cranial measurements. In fact, discriminant analysis revealed that the two species are more distinctive in the San Carlos Mountains than in Nuevo Leon, México, or Trans-Pecos, Texas.

Peromyscus pectoralis collinus was described originally by Hooper (1952) on the basis of specimens from the San Carlos Mountains. As currently understood, it also includes populations occurring in the Sierra de Tamaulipas and the tropical part of the eastern slopes of the Sierra Madre Oriental in Tamaulipas (Schmidly, 1972).

Peromyscus boylii ambiguus Alvarez

Brush Mouse (Ratón)

Specimens examined (204).—2.2 mi. W Rancho Carricitos, 10; 0.3 mi. SW Rancho Carricitos, 5; 0.3-0.5 mi. SW Rancho Carricitos, 57; 0.5 mi. SW Rancho Carricitos, 1; 0.6-1.0 mi. WSW Rancho Carricitos, 17; 1 mi. WSW Rancho Carricitos, 3; 1.1 mi. WSW Rancho Carricitos, 65; 2.8 mi. WSW Rancho Carricitos, 12; 2.9 mi. W Marmolejo, 5; 2.7 mi. W Marmolejo, 4; 3.4 mi. NNE San Carlos, 6; Marmolejo, 5 (UMMZ); La Vegonia, 14 (UMMZ).

The brush mouse was the most common rodent collected by us in the San Carlos Mountains. More than 98 per cent of the specimens trapped came from two habitats, the riparian and pine-oak forests (Table 5). The riparian-thorn woodland ecotone was the only other habitat type in which this species was trapped. Dice (1937:253-254) listed a series of 35 specimens under the name "*Peromyscus boylii levipes*" from the San Carlos Mountains. Actually, two taxa were represented in that series, with 14 of the 35 specimens being referrable to *P. boylii* and the remainder to *P. pectoralis* (Hooper, 1952). All of the specimens of *P. boylii* were from La Vegonia Mine, 2900 feet elevation, at the lower edge of the pine belt.

TABLE 6.—*Mean+1SD of external and cranial measurements for* P. pectoralis *and* P. boylii *from the San Carlos Mountains. An asterisk indicates a signifcance level of* P=0.01.

Character	P. pectoralis (N=41)	P. boylii (=43)	F-value
Total length	207.71+11.78	193.02+12.66	30.24*
Tail length	107.00+9.67	96.23+8.38	29.84*
Body length	100.90+6.49	96.42+7.62	8.40*
Ear length	18.71+1.76	17.02+1.74	19.41*
Skull length	27.71+0.70	27.09+0.69	16.82*
Maxillary toothrow length	4.13+0.16	4.56+0.19	122.86*
Mastoid Breadth	11.72+0.28	11.52+0.29	10.68*
Length rostrum	10.50+0.34	10.55+0.40	0.42
Length nasals	10.03+0.36	10.41+0.43	19.75*
Breadth rostrum	4.55+0.18	4.72+0.23	14.09*
Length interparietal	3.10+0.29	2.67+0.24	52.16*

Dates of collection of pregnant females and reproductive data are as follows: 9 January 1975, one with three 10-millimeter fetuses; 9 January 1976, one with four 5-millimeter fetuses and one with four embryos too small to measure. Of nine other females collected in this same month, three had swollen uteri and six did not. None of the females captured in May was pregnant, nor did any exhibit swollen uteri. Testicular lengths for nine adult males collected in January averaged 7.7 (range, 4-14); for 16 males taken in May, 5.7 (range, 2-9). These data suggest that reproductive activity for this species is different from that of *P. pectoralis*. The latter species was reproductively active in May but not in January, whereas the reverse appears to be true for *P. boylii*.

According to the most recent revision of this species by Schmidly (1973), *P. b. ambiguus* is the subspecies occuring in the San Carlos Mountains. External and cranial measurements of this subspecies are given in Table 6. *P. boylii* and *P. pectoralis* are similar in external size and appearance, and they can be distinguished from one another as described in the account of the latter.

Baiomys taylori taylori (Thomas)

Northern Pygmy Mouse (Ratón)

Specimens examined (3).—0.3 mi. SW Rancho Carricitos, 2; 2.6 mi. WNW San Carlos, 1.

We found this species to be relatively rare in the area as only three specimens, two from the riparian forest near Rancho Carricitos and one from the thorn woodland habitat near San Carlos, were obtained. Dice (1937) did not record the northern pygmy mouse from the San Carlos area.

The latest systematic review of pygmy mice was by Packard (1960), and the specimens from Tamaulipas are arranged according to his systematic findings. External and cranial measurements of two of our specimens are: total length, 105, 95; length of tail, 39, 38; length of hind foot, 12, 14; length of ear, 10, 9; greatest length of skull, 18.4, 17.9; zygomatic breadth,

9.8, 9.5; length of nasals, 6.4, 5.8; least interorbital constriction, 3.4, 3.1; mastoid breadth, 9.8, 8.6; length of maxillary toothrow, 2.8, 2.6; depth of skull, 6.7, 6.4.

Sigmodon hispidus berlandieri Baird

Hispid Cotton Rat (Rata grande)

Specimens examined (24).—Rancho Carricitos, 1; 0.3 mi. SW Rancho Carricitos, 1; 2.6 mi. WNW San Carlos, 3; Tamaulipeca, 14 (UMMZ); Mulato, 5 (UMMZ).

We obtained this species only in grassy areas within the riparian and thorn woodland habitats. It was not common as only eight individuals were captured. Dice (1937:254) found cotton rats to be common on the desert plains and valleys leading to the foothills. He secured specimens in a cattail thicket near El Mulato (thorn woodland habitat) and in a grassy area beside a dry wash at Tamaulipeca (thorn woodland). None of the specimens we collected evidenced reproductive activity, but Dice (1937:254) reported pregnant females in late July (two with eight fetuses, 20 and 18 mm. in crown-rump length; one with five, 33 mm. in length; and one with seven small embryos).

The subspecies *S. h. berlandieri* currently is regarded as occuring throughout northern and western Tamaulipas (Hall, 1981:738). External and cranial measurements of five specimens from the San Carlos area are: total length, 215, 229, 260, 202, 230; length of tail, 84, 109, 87, 89, 100; length of hind foot, 29, 30, 31, 29, 31; length of ear, 18, 22, 19, 18, 17; greatest length of skull, 31.4, 31.2, 34.0, 30.9, 31.6; zygomatic breadth, 18.0, 17.6, 19.3, 18.2, 18.0; length of nasals, 11.1, 11.5, 12.4, 10.8, 11.8; least interorbital constriction, 4.4, 4.6, 4.8, 4.4, 4.4; mastoid breadth, 13.2, 13.0, 14.8, 12.9, 13.0; length of maxillary toothrow, 5.6, 5.6, 5.4, 5.8, 5.6; depth of skull, 12.0, 12.0, 13.2, 11.9, 12.4.

Neotoma micropus micropus Baird

Southern Plains Woodrat (Rata)

Specimens examined (17).—2.6 mi. WNW San Carlos, 1600 ft., 2; Tamaulipeca, 4 (UMMZ); Mulato, 11 (UMMZ).

The only specimens we obtained of this species were trapped in brushy areas of the thorn woodland habitat near San Carlos. Dice (1937) reported woodrats as common on the desert plains and in the lower mountains. He secured specimens near El Mulato in xeric thorn scrub and thorn woodland habitat, and at Tamaulipeca in xeric thorn scurb and deciduous thicket habitat.

Neither of our specimens, which were taken on 4 January, was in reproductive condition, but Dice (1937) collected pregnant females on 24 July (two fetuses approximately 36 mm. long) and 16 August (two fetuses, 34 and 39).

We follow Birney (1973) in assigning our specimens to *N. m. micropus*. Average external and cranial measurements (extremes in parentheses) of eight specimens from the San Carlos region are: total length, 347 (339-362); length of tail, 147 (134-158); length of hind foot, 34.6 (34-36); length of ear, 26.4 (24-28); greatest length of skull, 44.4 (43.0-44.8); zygomatic breadth, 23.2 (22.0-25.3); length of nasals, 16.2 (15.2-17.8); least interorbtial constriction, 5.7 (5.4-6.0); length of maxillary toothrow, 8.2 (7.5-8.8); depth of skull, 16.1 (15.7-17.5).

Mus musculus brevirostris Waterhouse

House Mouse (Ratón)

Specimens examined (4).—0.3 mi. SW Rancho Carricitos, 3; 8.3 mi. SSE San Carlos, 1.

The only specimens of the house mouse we obtained were collected in houses at Rancho Carricitos and Rancho San Francisco. A female collected on 6 January at Rancho Carricitos had eight 19-millimeter fetuses. A lactating female, with a swollen uterus and six placental scars (three per horn), was trapped on 26 March at Rancho San Francisco.

We follow Schwarz and Schwarz (1943) in assigning these to *M. m. brevirostris*, which evidently ranges from South America to the southern United States. External and cranial measurements of three of our specimens are: total length, 168, 160, 166; length of tail, 78, 69, 76; length of hind foot, 19, 18, 17; length of ear, 14, 13, 15; greatest length of skull, 21.8, 21.5, 22.8; zygomatic breadth, 11.4, 11.5, 11.7; length of nasals, 6.9, 7.2, 8.2; least interorbital constriction, 3.6, 2.9, 3.4; mastoid breadth, 9.8, 9.4, 9.9; length of maxillary toothrow, 3.0, 3.2, 3.3; depth of skull, 7.9, 7.4, 8.6.

Rattus rattus (Linnaeus)

Black Rat (Rata)

Specimen examined (1).—Rancho Carricitos, 1900 ft. 1.

Our only specimen is a juvenile taken on 6 January in a house at Rancho Carricitos. Undoubtedly, black rats also occur in dwellings in other villages in the mountains. We made no attempt to allocate our specimen to a definite subspecies, because this is usually impossible (Hall, 1981:838).

Canis latrans microdon Merriam

Coyote (Coyote)

Specimens examined (4).—0.3-0.5 mi. SW Rancho Carricitos, 1; San José, 1 (UMMZ); Mulato, 2 (UMMZ).

This is probably one of the more abundant and most significant predators occurring within the San Carlos region. We heard coyotes at night during our work in the area, but the only specimen obtained was purchased at Rancho Carricitos. Dice (1937:251) secured specimens from San Miguel, which is only a short distance from Tamaulipeca, and from El Mulato.

We have assigned our specimens to *C. l. microdon* because Alvarez (1963:454-455) assigned specimens from nearby localities in Tamaulipas to this subspecies. External and cranial measurements of specimens (two females, one male) are as follows: total length, 1050, 1055, 1130; length of tail, 307, 297, 329; length of hind foot, 178, 188, 188; length of ear, 103, 100, 107; condylobasilar length, 163.4, 167.5, 179.8; zygomatic breadth, 88.0, 88.8, 97.2; width of braincase, 53.6, 52.9, 56.1; postorbital width, 31.6, 32.9, 33.4; interorbital width, 27.6, 29.5, 30.4; paletal width across Ml, 50.6, 51.4, 54.6; width between auditory bullae, 9.8, 12.0, 13.2.

Urocyon cinereoargenteus scotti Mearns

Gray Fox (Zorro)

Specimens examined (8).—0.3-0.5 mi. SW Rancho Carricitos, 3; Marmolejo, 2 (UMMZ); Mulato, 3 (UMMZ).

This is evidently one of the more abundant carnivores occurring in the San Carlos Mountains, where it seems to prefer wooded or canyon situations. We captured an adult male gray fox (testis = 22) on 13 January in a steel trap baited with rodent carcasses in the riparian habitat near Rancho Carricitos. We also purchased two skins from residents of Rancho Carricitos, but we were unable to determine precisely from whence these came. Dice (1937:250) secured specimens from inhabitants of Marmolejo and El Mulato and saw several foxes alive on the limestone hills (deciduous thickets) near Tamaulipeca.

As currently understood, gray foxes from the San Carlos Mountains are referred to the subspecies *U. c. scotti*, which occurs widely in the southwestern United States and northern México (Hall, 1981:944). External and cranial measurements of two adult specimens from the San Carlos area are: total length, 840, 1015; length of tail, 369, 363; length of hind foot, 132, 140; length of ear, 68, 79; condylobasilar length, 119.9, 119.5; zygomatic breadth, 57.4, 65.4; width of braincase, 43.3, 45.2; postorbital width, 28.6, 24.6; interorbital width, 19.4, 24.7; width across Ml, 32.4, 33.4; width between auditory bullae, 8.4, 8.2.

Bassariscus astutus flavus Rhoads

Ringtail (Pintorabo)

Specimens examined (18).—2.6 mi. WNW San Carlos, 1; 0.3-0.5 mi. SW Rancho Carricitos, 14; 0.3 mi. SW Rancho Carricitos, 3.

The ringtail is probably quite common in the San Carlos Mountains, but it is seldom seen because it frequents rocky, inaccessible habitats. All of our specimens were purchased from natives, who told us that ringtails are common in the rock cliffs of the foothills and lower mountains where riparian and thorn woodland habitats predominate.

We have assigned our specimens to *B. a. flavus*, which occupies a wide geographic range in Texas, Oklahoma, New Mexico, Colorado, and northeastern México (Hall, 1981:964). Average external measurements (ranges in parentheses) of six adult specimens from the San Carlos Mountains are: total length, 672.2 (630-707); length of tail, 316.8 (264-354); length of hind foot, 63.2 (58-69); length of ear, 44.8 (43-47).

Procyon lotor fuscipes Mearns
Raccoon (Tejón)

Specimens examined (3).—Rancho Carricitos, 1; Marmolejo, 2 (UMMZ).

According to Dice (1937:249), the raccoon is of general occurrence over the plains and the valleys in the mountains, and over at least the lower hills. We found it to be relatively rare in the riparian community along Arroyo Ebanal, although we would expect it anywhere in the San Carlos region where sources of water are associated with wooded areas. Long-term residents told us that raccoons are widely distributed but uncommon in the area. The only specimen we obtained was a nonpregnant adult female purchased on 12 January 1976. On 10 January, Hendricks saw a raccoon in the riparian habitat near our campsite at Rancho Carricitos. Dice (1937) purchased an adult female raccoon and several young at Marmolejo.

We have assigned our specimens to *P. l. fuscipes* on geographic grounds (Hall, 1981:971). Goldman (1950:51) assigned a specimen from Marmolejo to this subspecies; Alvarez (1963:457) allocated a specimen from Santa Rosa, which is about 50 mi. SW San Carlos, to *P. l. hernandezii*. Our specimens show features characteristic of both *fuscipes* and *hernandezii*, but overall they are more like the former than the latter. They are dark grayish in color, with small, black postauricular spots; the braincase is high, with a moderately broad frontal region and weakly developed postorbital processes. Based on this evidence, it seems likely that the San Carlos Mountains represent an area of intergradation between *fuscipes* and *hernandezii*. External and cranial measurements of two specimens from the San Carlos area are: total length, 805, 820; length of tail, 265, 285: length of hind foot, 121, 120; length of ear, 65, 63; condylobasilar length, 110.1, 115.4; zygomatic breadth, 69.8, 67.6; width of braincase, 48.2, 50.4; postorbital width, 23.2, 23.4; interorbital width, 23.4, 21.4; width across M1, 37.0, 43.2; width between auditory bullae, 14.0, 16.8.

Nasua narica molaris Merriam
Coati (Mapache)

Specimens examined (5).—8 km. N San Carlos, 1; San José, 4 (UMMZ).

We purchased a specimen of this species from a native who said he shot it north of San Carlos on the road to San Nicolas. The countryside in this area consists of numerous rolling hills dissected with rocky outcrops and covered with scrub-oak thickets. Dice (1937:249) reported specimens from San José, El Mulato, and Marmolejo, and he suspected that the coati occurs both on the plains and in the foothills but not in the forests of the mountains. Local residents, however, told us the coati occurs throughout the mountainous portions of the San Carlos area.

We follow Hall (1981:976) in assigning our specimens to the subspecies *N. n. molaris* on geographical grounds.

Mustela frenata frenata Lichtenstein

Long-tailed Weasel (Hurón)

Specimens examined (2).—Rancho Carricitos, 1; 0.9 mi. WSW Rancho Carricitos, 1.

Two specimens of this species were obtained during our work, and another individual was sighted. Mr. Bret Marsh and Ms. Pam Bilderback saw a long-tailed weasel running along a pathway bordering Arroyo Ebanal at 8:35 PM on 5 January 1976. A skin was purchased at Rancho Carricitos on 10 January 1976, but we could not determine exactly where this animal had been obtained. On 13 January 1976 an adult male, with testis 12 mm. in length, was taken in a steel trap set at the face of a cliff near the bottom of Arroyo Ebanal. These are the first documented occurrences of this species in the San Carlos Mountains. Natives reported an animal, the description of which seemed to apply to a large weasel of some kind, that apparently lived about the thorn fences bordering the cultivated fields at El Milagro, but no specimen was secured (Dice, 1937:250).

Two subspecies of the long-tailed weasel have been reported from Tamaulipas; *M. f. frenata*, which occurs in the central and northern parts of the state, and *M. f. tropicalis*, which occurs in the tropical area in the southern part of the state (Alvarez, 1963:459). Our specimen, based on the descriptions of these two subspecies by Hall (1951a:341), clearly is referable to *M. f. frenata*. The blackish coloration of the head extends posteriorly on the top and sides of the neck almost halfway to the shoulders, the black tip of the tail is one-fourth the length of the tail and about the same length as the hind foot, and the postorbital breadth of the skull is less than the length of upper molar and premolar toothrow. External and cranial measurements of our specimen are: total length, 450; length of tail, 182; length of hind foot, 49; length of ear, 23; greatest length of skull, 53.6; zygomatic breadth, 29.8; mastoid breadth, 25.9; least interorbital constriction, 8.2; length of maxillary toothrow, 14.8; depth of skull, 19.2.

Taxidea taxus berlandieri Baird

Badger (Talcoyote)

Specimen examined (1).—12 mi. NW San Carlos, 1300 ft., 1 (KU).

Alvarez (1963:459-460) reported the only record of the badger from the vicinity of San Carlos. An adult male (testis 34 in length) was taken in a steel-trap baited with a bird body and rabbit meat and set in front of a hole in the thorn woodland habitat 12 mi. NW San Carlos. We did not observe any evidence of badgers during our work in the area nor did Dice (1937) when he was in the mountains. However, the species is undoubtedly of rare occurrence on the lower plains and valleys leading into the foothills where ground squirrels and other burrowing rodents are present. Badgers are probably absent from the timbered and rocky mountain areas where there are no burrowing rodents.

We have followed Long (1972:749) in assigning our specimen to *T. t. berlandieri.* Its external and cranial measurements are: total length, 710; length of tail, 115; length of hind foot, 110; length of ear, 55; condylobasal length, 123.1; zygomatic breadth, 81.1; mastoid breadth, 75.5; interorbital constriction, 29.3; least postorbital constriction, 27.6; length of maxillary toothrow, 42.7.

Spilogale putorius interrupta (Rafinesque)

Eastern Spotted Skunk (Zorilla)

Specimen examined (1).—Tinaja, 1.

The only specimen available from the San Carlos Mountains is a female obtained on 11 January 1976 near human dwellings at Tinaja in thorn woodland habitat. Dice (1937) did not obtain this species when he worked in the area. Natives told us that spotted skunks occur throughout the thorn woodlands around Gavilan and Marmolejo, and that they are occasionally seen around arroyos near Rancho San Francisco.

Our specimen is assigned to the subspecies *interrupta* following Van Gelder (1959:270-279), who regarded specimens from Tamaulipas as intergrades between *S. p. leucoparia* and *S. p. interrupta.* External and cranial measurements of our specimen are: total length, 365; length of tail, 125; length of hind foot, 40; length of ear, 28; greatest length of skull, 50.4; condylobasilar length, 50.0; zygomatic breadth, 30.8; postorbital constriction, 14.2; least interorbital constriction, 14.4; mastoid breadth, 27.9; length of maxillary toothrow, 16.6.

Mephitis mephitis varians Gray

Striped Skunk (Zorilla)

Specimens examined (13).—0.3-0.5 mi. SW Rancho Carricitos, 1; 0.3 mi. SW Rancho Carricitos, 3; 2.6 mi. WNW San Carlos, 1; San José, 2 (UMMZ); Marmolejo, 5 (UMMZ).

We obtained striped skunks from the riparian habitat near Rancho Carricitos and from the thorn woodland habitat near San Carlos. Also, long-term residents of the area told us they are common around towns and houses in both the foothills and xeric thorn scrub habitats. Dice (1937:250) reported this species from San José (thorn woodland), El Mulato (xeric thorn scurb), and Marmolejo (deciduous thickets). One of our specimens, a male, had testes that measured 23 in length when taken on 14 January.

M. m. varians occurs from the Central Great Plains to northern México (Hall, 1981:1021), and this is the name applied by Alvarez (1963:461) to specimens from the San Carlos area. External and cranial measurements of four of our specimens are: total length, 638, 691, 735, 621; length of tail, 308, 300, 360, 360; length of hind foot, 65, 68, 63, 70; length of ear, 21, 18, 26, 36; greatest length of skull, 71.6, 68.8, 70.0, 64.8; condylobasilar length, 68.8, 67.0, 69.8, 63.4; zygomatic breadth, 40.3, 42.4, 42.4, 40.8; postorbital constriction, 17.0, 18.6, 18.0, 18.4; least interorbital constriction, 19.2, 19.8, 19.4, 18.6; mastoid breadth, 34.0, 34.6, 34.6, 32.6; length of maxillary toothrow, 23.1, 21.5, 23.2, 21.2.

Conepatus leuconotus texensis Merriam
Eastern Hog-nosed Skunk (Zorilla)

Specimens examined (4).—Rancho Carricitos, 3; Mulato, 1 (UMMZ).

Three male specimens of this species were purchased on 10 January 1976 from natives who reported they were taken in the vicinity of Rancho Carricitos. Testicular length of two of the males was 20 mm. each. On 8 January 1975, while camped in the pine-oak forest above Rancho Carricitos, a skunk of this species attacked several students and was subsequently shot. The animal was taken to Brownsville and turned in to U.S. Public Health officials who, after conducting tests, reported it was rabid. Dice (1937:250) obtained a single individual of this species at El Mulato, which is on the Tamaulipan Plain surrounding the San Carlos Mountains.

We follow Hall (1981:1026) in assigning our specimen to the subspecies *C. l. texensis* on geographical grounds. External and cranial measurements of an adult male from Mulato are: total length, 760; length of tail, 297; length of hind foot, 84; length of ear, 29; greatest length of skull, 87.1; zygomatic breadth, 54.0; postorbital constriction, 24.0; least interorbital constriction, 27.2; mastoid breadth, 47.2; length of maxillary toothrow, 25.6.

Felis concolor Linnaeus
Mountain Lion (Léon)

Specimens examined.—None.

We know of no scientific specimens of the mountain lion from the San Carlos Mountains. However, one of us (Hendricks) saw a cougar on the evening of 10 January 1975 while walking along a pathway near Arroyo Ebanal in the vicinity of Rancho Carricitos. Also, Hendricks saw a fresh, wet skin of a mountain lion on 27 September 1975 in the village of San Carlos. The man in possession of the skin informed Hendricks that he had killed it near San Nicolas a few days earlier. Long-term residents of the area told us that mountain lions are regularly seen and shot in the mountains and surrounding xeric areas. Dice (1937) was unable to document the occurrence of the cougar in the San Carlos Mountains, other than by word-of-mouth from residents familiar with the area.

Felis pardalis albescens Pucheran
Ocelot (Tigrillo)

Specimens examined (2).—San José, 1 (UMMZ); Mulato, 1 (UMMZ).

Dice (1937:251) secured specimens of the ocelot from residents of El Mulato and San José. He reported that ocleots were rare, but generally distributed in the several life belts of the San Carlos region. We obtained no evidence of occurrence during our work, but natives of the area told us that ocelots are seen occasionally in the mountains.

Alvarez (1963:464) referred a specimen from 10 mi. N Altamira to the subspecies *albescens* on the basis of its small size, and he concluded that all ocelots from Tamaulipas should be referred to that subspecies. We have followed this arrangement.

Felis rufus texensis (J. A. Allen)
Bobcat (Gato montés)

Specimens examined (2).—0.3-0.5 mi. SW Rancho Carricitos, 1; Mulato, 1 (UMMZ).

The bobcat is reported by residents of the area to be relatively common both in the mountains and on the plains. Only two specimens are available from the San Carlos Mountains, one secured from natives of El Mulato by Dice (1937:251) and another purchased by us at Rancho Carricitos.

We follow Anderson (1972:386-387) in use of the generic name *Felis* for bobcats previously known under the name *Lynx*. The subspecific name *F. r. texensis* has been applied to bobcats from northern Tamaulipas (Hall, 1981:1053), and we have followed this arrangement.

Tayassu tajacu angulatus (Cope)
Collared Peccary (Jabalí)

Specimen examined (1).—Cruillas, 1 (UMMZ).

According to Dice (1937:256), peccaries occur everywhere in the San Carlos region except in the higher mountains. He reported that they were especially numerous at Tamaulipeca, and he collected a fragment of a jaw near El Milagro. We did not obtain any specimens of the peccary, nor did we observe any sign of it during our work in the region, although residents of the area told us they are common in the canyons and lower hills of the mountains.

We follow Wetzel (1977:7) in the use of the generic name *Tayassu* for collared peccaries previously known under the name *Dicotyles* (Woodburn, 1968). The subspecific name *T. t. angulatus* has been applied to peccaries throughout Tamaulipas (Hall, 1981:1081), and we have followed this arrangement.

Odocoileus virginianus texanus (Mearns)

White-tailed Deer (Venado)

Specimen examined (1).—Mulato, 1 (UMMZ).

We did not obtain any specimens of white-tailed deer from the San Carlos Mountains, although they are reported by residents to be numerous in the foothills and mountainous areas. Dice (1937:256) purchased a set of antlers at El Mulato and saw an adult deer at Sardinia Pass. The owner of Rancho San Francisco told us that these animals congregate at the water tanks on various ranches in the area.

Dice (1937:256) assigned the white-tailed deer in the San Carlos region to the subspecies *O. v. texanus* on the basis of the antlers obtained at El Mulato. We have followed this arrangement.

DISCUSSION

Our field work and survey of the literature indicate that 56 species of mammals occur, or have recently occurred, in the area of the San Carlos Mountains. Of these, there are a marsupial, an edentate, a perissodactyl, an artiodactyl, two insectivores, three rabbits, 13 carnivores, 15 rodents, and 19 bats. Another seven species possibly occur or possibly have occured in the region. Their status may be summarized as follows:

Onychomys leucogaster.—Although we did not collect the northern grasshopper mouse in the San Carlos Mountains, it has been taken in Cd. Victoria and supposedly occurs over all of northern Tamaulipas (Alvarez, 1963:448). We would expect this species in the grassland situations interspersed in the thorn woodland habitats of the San Carlos Mountains.

Canis lupus.—No positive evidence of the presence of the gray wolf in the San Carlos Mountains has been found. On the major distribution maps of this species published by Leopold (1959:400; Hall, 1981:932), San Carlos is situated at the edge of the region in which the wolf is considered to be extinct in Tamaulipas.

Ursus americanus.—We know of no specific scientific specimens of the black bear from the San Carlos Mountains, although the species undoubtedly occurred there in the past and probably still occurs in limited numbers. Dice (1937) did not report bears in the mountains, but natives familiar with the area reported seeing bear bothering livestock at Rancho Rosario (above Marmolejo) several years ago, and these same individuals told us they saw the footprints of a bear at the same ranch in 1979. The only verified records of black bear in Tamaulipas are from Agua Linda (Goodwin, 1954:14) and La Joya de Salas (specimen number 102430 UMMZ) in the high and remote parts of the Sierrra Madre Oriental in the southern part of the state.

Felis onca.—Dice (1937:251) mentioned the possible sign of a jaguar near San José and reported that natives told him these cats lived mostly in the higher parts of the mountains and wandered widely over the area. The only localities in Tamaulipas from which specimens have been documented are the Gomez Farias region and the Sierra de Tamaulipas in the southern part of the state (Alvarez, 1963:463).

Felis wiedii.—The occurence of this cat cannot be documented in the region of study, but it is known from southern Texas (Eagle Pass, Maverick County—Goldman, 1943) and southern Tamaulipas (Rancho del Cielo—Goodwin, 1954:15) and, hence, likely does occur, or has occurred historically, in the San Carlos Mountains.

Felis yagouaroundi.—Although we obtained no specimens nor made sightings of this species, the jaguaroundi is known from places to the north (Matamoros—Hall, 1981:1049) and south (Gomez Farías—Goodwin, 1954:15) of the San Carlos Mountains, and it likely has occurred historically, or still occurs, in the area. Jaguaroundis frequent chaparral and dense forests, seemingly having preference for localities near streams. Similar situations in the San Carlos region should provide suitable habitat for this species, and residents of the area tell us that this cat is seen occasionally in the hills and mountains.

Antilocapra americana.—The pronghorn is extinct in Tamaulipas, but it has occured in the northern part of the state and perhaps in the San Carlos region in relatively recent time (more than 125 years ago—Alvarez, 1963:467). It probably was eliminated by hunting pressure and habitat alteration.

Mammalian Distribution within the San Carlos Mountains

We recognize five major mammalian distributional zones in the San Carlos Mountains. These correspond closely with the major vegetational types and the distribution of some indicator species of mammals. Good indicator species are relatively abundant and are restricted, or nearly so, to the zone for which they are an indicator. The five zones and their indicator species of mammals are: xeric thorn scrub—*Perognathus flavus, Sylvilagus audubonii, Lepus californicus*; thorn woodland—*Spermophilus mexicanus, Perognathus hispidus, Spilogale putorius*; deciduous thickets—*Peromyscus pectoralis*; riparian forests—*Sciurus aureogaster, Peromyscus leucopus*; pine-oak forests—*Peromyscus boylii, Sciurus alleni*.

The distribution of each species within the major mammalian distributional zones is given in Appendix 1. The riparian habitat contains the most mammalian species; the deciduous thickets, the fewest. The xeric thorn scrub and pine-oak forest habitats contain relatively few species, whereas the thorn woodland supports an intermediate number. The number of mammalian species present in each zone, shared between zones, and the Burt Coefficients of Similarity between zones is given in Table 7. The xeric thorn scrub has a relatively low coefficient with all habitats. The highest coefficients are between thorn woodland-deciduous thickets (0.548) and riparian forest-pine-oak forest habitats (0.535). The riparian forest shares a

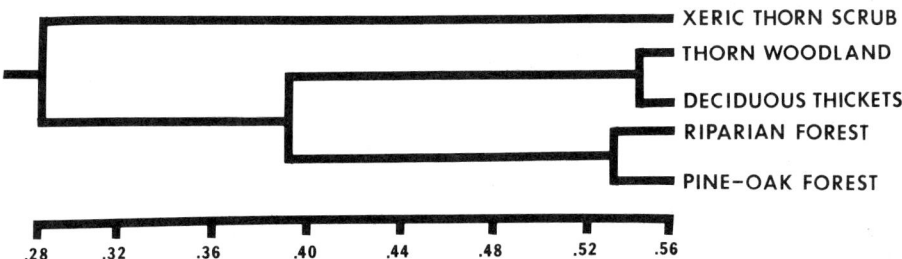

FIG.8.—Phenogram resulting from the clustering (Unweighted Pair Group Method Using Arithmetic Averages, UPGMA) of Burt Coefficients of Similarity among mammalian distribution zones as given in Table 7.

larger number of species with the thorn woodland (22), deciduous thickets (17), and pine-oak forest zones (23). Cluster analysis of the similarity coefficients (Fig. 8) places xeric thorn scrub in a separate branch of the phenogram far apart from the other habitat types. The deciduous thickets-thorn woodland and riparian forest-pine-oak forest habitats form pair-wise but separate groups in the other main branch of the phenogram.

These results indicate that the xeric thorn scrub fauna, which is characteristic of the desertlike areas surrounding the mountains, is quite distinct from that of the foothills and montane habitats. The large number of species shared between the riparian forest and the other mammalian zones is reasonable inasmuch as the riparian forest extends through the major habitat zones from the mountain tops to the deserts. The high similarity between the thorn woodland and the deciduous thickets is understandable because these habitats usually are in proximity to one another.

Affinities and Derivation of the Mammalian Fauna

We compared the mammalian fauna of the San Carlos Mountains with that of other specific areas for which the mammalian fauna is relatively well known. The following areas, within the same geographic region, were chosen (references in parentheses are those used to develop the faunal list

TABLE 7.—*Similarity of mammalian faunas occuring in the five mammalian distribution zones recognized in the San Carlos Mountains, Tamaulipas. The numbers above the diagonal represent the Burt Coefficients of Similarity between the zones, whereas the number below the diagonal represent the number of species shared between areas.*

	Xeric Thorn Scrub	Thorn Woodland	Deciduous Thickets	Riparian Forest	Pine-Oak Forest
Xeric Thorn Scrub		0.351	0.357	0.231	0.235
Thorn Woodland	13		0.548	0.423	0.368
Deciduous Thickets	10	17		0.378	0.428
Riparian Forest	12	22	17		0.535
Pine-Oak Forest	8	14	12	23	

TABLE 8.—*Similarity of the total mammalian faunas occuring in seven selected geographic areas of Texas, Nuevo Leon, and Tamaulipas. The numbers above the diagonal represent the number of species shared between areas, whereas the numbers below the diagonal represent the Burt Coefficients of Similarity between areas.*

	San Carlos	Monterrey	Cd. Victoria	Gomez Farias	Sierra de Tamaulipas	Brownsville	Nuevo Leon
San Carlos		46	43	31	36	38	35
Monterrey	0.638		41	30	31	35	30
Cd. Victoria	0.597	0.512		41	37	35	30
Gomez Farias	0.322	0.291	0.461		41	35	22
Sierra de Tamaulipas	0.514	0.383	0.514	0.512		29	24
Brownsville	0.567	0.460	0.479	0.412	0.414		36
Nuevo Leon	0.555	0.405	0.422	0.242	0.353	0.654	

given in Appendix 2): Laredo, Texas, and Nuevo Laredo, Tamaulipas (Blair, 1952; Alvarez, 1963; Davis, 1974; specimens in the collection at Texas A&M University); Brownsville, Texas, and Matamoros, Tamaulipas (Blair, 1952; Alvarez, 1963; Davis, 1974); Monterrey, Nuevo León (Koestner, 1938; Koopman and Martin, 1959; Hall, 1981; Villa-R., 1967; Guzman, 1968); Cd. Victoria, Tamaulipas (Koopman and Martin, 1959; Alvarez, 1963); Sierra de Tamaulipas, Tamaulipas (Hooper, 1953; Martin *et al.*, 1954; Alvarez, 1963; specimens in the collection at Texas A&M University); and the Gomez Farías region, Tamaulipas (Baker, 1951; Goodwin, 1954; Koopman and Martin, 1959; Alvarez, 1963).

Laredo-Nuevo Laredo and Brownsville-Matamoros are major urban sites in the Tamaulipan Plain region along the Texas-Mexican border. The vegetation is characterized by a more or less dense growth of thorny shrubs with a few wet, marshy areas near the coast. Monterrey and Cd. Victoria are major cities situated in the Sierra Madre Oriental of Nuevo León and Tamaulipas, respectively. The vegetation in these areas grades from a coastal plain thorn scrub at the base of the mountains to a dry pine-oak forest and a humid pine-oak forest at higher elevations. Gomez Farías lies in southern Tamaulipas in the tropical part of the Sierra Madre Oriental, immediately south of the Tropic of Cancer. In addition to humid and dry pine-oak forests, this area also contains tropical deciduous and cloud forest habitats. The Sierra de Tamaulipas, like the Sierra San Carlos, represent an isolated mountain range situated immediately east of Gomez Farías in the Tamaulipan coastal plain; habitat types include tropical thorn forest, tropical deciduous forest, montane scrub, and pine-oak forests (Martin *et al.*, 1954). The most striking biogeographic feature of this region of northeastern México is the faunal "break" in the vicinity of Gomez Farías in southern Tamaulipas, corresponding to the northern limit of tropical deciduous forest vegetation (Martin, 1958; Koopman and Martin, 1959).

The total mammalian fauna of the San Carlos Mountains shows the highest similarity with the fauna of Monterrey and Cd. Victoria (46 and 43 species, respectively, are shared between these areas and San Carlos—see Table 8). Intermediate similarity values were obtained with Brownsville, Laredo, and the Sierra de Tamaulipas, and the lowest similarity value was found to be with the mammalian fauna at Gomez Farías. A clustering of the similarity coefficients is shown in Fig. 9.

We also compared these same geographic areas using only bat and rodent faunas. This changes the relationships among the areas somewhat, chiefly by changing the position of Victoria for the bats and the position of Brownsville-Laredo for rodents (see Fig. 9). The bat fauna of San Carlos shows the highest similarity to that of Monterrey (16 shared species). Rodents may be somewhat better indicators of faunal resemblance because they are less vagile, are highly affected by the environment, and generally do not have widespread geographic ranges (Genoways et al., 1979). The rodent fauna of San Carlos shows the highest similarity to the rodent fauna of Cd. Victoria (13 shared species).

These results indicate that the mammalian fauna of San Carlos should be considered more closely related to that of Monterrey and Cd. Victoria, which lie at the edge of the Sierra Madre Oriental, than to Brownsville and Laredo in the Tamaulipan Plain or to Gomez Farías and the Sierra de Tamaulipas in the tropical portion of southern Tamaulipas. The vegetation types in the San Carlos Mountains are climatically equivalent to the coastal-plain scrub, piedmont scurb, and montane low forest vegation types (described by Müller, 1939) along the lower, eastern slopes of the Sierra Madre Oriental. In view of this physiognomic and climatic similarity, it is not surprising that the coastal plain and Sierra San Carlos exhibit a close faunal relationsip to the Monterrey and Cd. Victoria regions of the Sierra Madre.

Recent mammals that occur in the San Carlos Mountains include a mixture of species from several different faunal areas including: 1) southern species characteristic of lowland tropical and subtropical habitats; 2) interior species charcteristic of the southwest or arid Mexican Plateau; 3) grassland (steppe) species charcteristic of the Great Plains; 4) trans-plateau species characteristic of the dry pine-oak forest and open woodlands of the Mexican Highland and surrounding the Mexican Plateau; and 5) deciduous forest species characteristic of the deciduous forests of the eastern United States.

The species in this mixed assemblage of mammals probably reached the San Carlos region via one of three main dispersal routes. These include 1) a northern route for species from the Great Plains, deciduous forests of the eastern United States, and the southwest to reach the San Carlos region via Texas, northeastern Coahuila, and Nuevo León; 2) a southern route for mammals of the Gulf Coast; and 3) a montane route for mammals of dry, open woodland habitats to enter the San Carlos region from the south and west.

Species occurring in the San Carlos region that have faunal affinities with the Great Plains grasslands are as follows: *Spermophilus mexicanus*, *Perognathus hispidus*, *Neotoma micropus*, and *Spilogale putorius*. Those

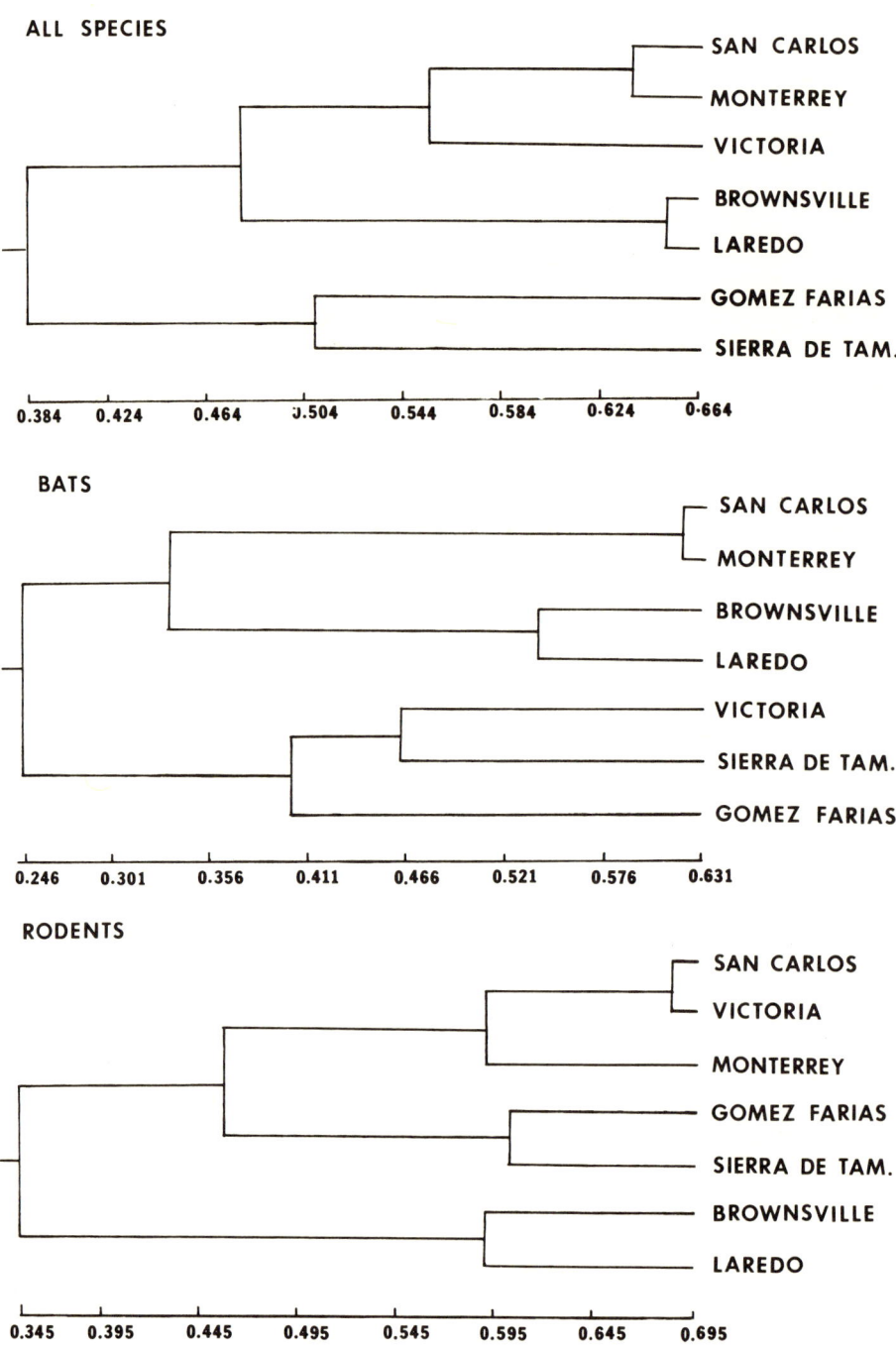

Fig. 9.—Phenogram resulting from the clustering (UPGMA) of Burt Coefficients of Similarity among the total mammalian faunas (top), bat faunas (middle), and rodent faunas (bottom) of selected geographic areas in South Texas and northeastern México.

associated with eastern deciduous forests of the United States include: *Cryptotis parva, Nycticeius humeralis, Sylvilagus floridanus*, and *Peromyscus leucopus*.

Species occurring at San Carlos having faunal affinities with the arid, interior Mexican Plateau include: *Notiosorex crawfordi, Choeronycteris mexicana, Myotis velifer, Myotis auriculus, Myotis californicus, Pipistrellus hesperus, Plecotus townsendii, Antrozous pallidus, Sylvilagus audubonii, Lepus californicus, Perognathus flavus, Peromyscus pectoralis*, and *Bassariscus astustus*. The mammals in this group were able to penetrate the Sierra del Carmen-Sierra Madre Oriental Axis Filter Barrier (described by Baker, 1956), which forms the eastern boundary of the Mexican Plateau in Coahuila and Nuevo Léon.

Many Trans-Plateau species charcteristic of the dry, oak-pine and savannah grassland habitats in the southern part of the Mesa Central or the Highlands of southern México also occupy tropical or subtropical habitats along the Gulf Coast of México. Mammals in this group that occur at San Carlos include: *Leptonycteris nivalis, Lasiurus ega, Lasiurus intermedius, Sciurus aureogaster, Liomys irroratus, Sigmodon hispidus, Reithrodontomys fulvescens, Baiomys taylori*, and *Peromyscus boylii*.

Species of the Sierra Madre Oriental that are confined to sub-boreal and boreal habitats tend not to occur in the San Carlos Mountains. Few species are restricted to the montane (pine-oak) habitat in the San Carlos Mountains, and the faunal affinities of most of these mammals are with the lowlands more than with the montane habitats of the Sierra Madre. Rodents and shrews in particular illustrate this affinity. Throughout the Sierra Madre Oriental in Nuevo León certain rodents and shrews can be expected wherever pine-oak forests are found; some of those species apparently favor arid pine-oak woodlands (low montane forest) of lower elevations, whereas others inhabit the humid or montane mesic forest of higher elevations (Table 9). No species characteristic of the montane humid forest in the Sierra Madre occurs at San Carlos, and only two species of the low montane forest occur in pine-oak forest of the Sierra San Carlos. One of these, *Sciurus alleni*, is restricted in distribution to the Sierra Madre and Sierra San Carlos in Coahuila, Nuevo León, and Tamaulipas. The other, *Peromyscus boylii*, is widespread throughout the pine-oak forests of southern and western México.

Reasons for the absence of a distinct pine-oak faunal component among the mammals of the Sierra San Carlos probably relates to two factors, the small size of the habitat and a postglacial thermal maximum, which Martin *et al.* (1954) hypothesized to account for a similar charcteristic in the faunal composition of the Sierra de Tamaulipas. The pine-oak formation of the Sierra San Carlos occupies a small area, covering less than 700 square kilometers. Optimal population conditions normally are not realized by animals restricted to small isolated habitats, and the possibility of local extermination of species confined to the pine-oak formation, therefore,

TABLE 9.—*Occurrence of montane shrews and rodents in the Sierra Madre Oriental and Sierra San Carlos.*

Species	Sierra Madre Oriental		Sierra San Carlos
	Montane Low Forest	Montane Humid Forest	
Sorex milleri		+	
Sorex saussurei		+	
Eutamias bulleri		+	
Sciurus alleni	+		+
Spermophilus variegatus	+		
Thomomy bottae	+	+	
Pappogeomys castanops (subspecies planifrons)	+	+	
Reithrodontomys megalotis	+		
Peromyscus maniculatus	+	+	
Peromyscus boylii	+		+
Peromyscus difficilis	+		
Peromyscus melanotis		+	
Neotoma albigula	+		
Neotoma mexicana		+	
Microtus mexicanus		+	

probably is much greater than for the same species inhabiting the extensive pine-oak belt of the Sierra Madre Oriental. Recent xerothermic conditions in northeastern México probably caused a further restriction of the pine-oak formation to isolated peaks and north-facing ravines, subdividing the habitat into numerous isolated pockets. Under such conditions, animals confined to the pine-oak belt would be in danger both of increased competition from invading lowland species and local extirpation in an environment too small to support adequate breeding populations.

The tropical element in the San Carlos fauna is evidenced by the seven species characteristic of the lowland tropical fauna (as defined by Koopman and Martin, 1959), and modified by us) that occur in the region. Those include: *Didelphis virginiana*, *Mormoops megalophylla*, *Desmodus rotundus*, *Felis pardalis*, *Nasua narica*, *Dasypus novemcinctus*, and *Tayassu tajacu*. The vampire bat (*Desmodus rotundus*) is the only species in this group with a distribution that corresponds roughly to the boundary of the warm tropical lowlands; the other species either range into montane habitats or reach temperate latitudes.

The Gulf lowlands of northeastern México and southern Texas represent an area of broad faunal transition (that is, a faunal gradient) rather than a distinctive faunal unit. Only one mammal, the eastern hog-nosed skunk (*Conepatus leuconotus*), is endemic to this region, and it occurs in the San Carlos Mountains. Beginning in southern Tamaulipas and continuing to the north toward southern Texas and Nuevo Léon, a strong vegetational gradient is evident. It starts with the tropical deciduous forest (which is well developed in the Sierra de Tamaulipas and Sierra Madre Oriental of southern Tamaulipas) and grades into rather barren thorn forest and thorn scrub habitats as the Rio Grande Valley is approached. Across this gradient a rapid depletion of tropical forms occurs, because 17 mammalian genera reach limits of known range between 23° and 24° north latitude in southern Tamaulipas (Koopman and Martin, 1959).

Several widespread North American or Pan-American elements occur in the mammalian fauna of the San Carlos Mountains. These species have broad geographic ranges and are of little value in determing the relationships of a fauna. The following species may be included in this category: *Lasionycteris noctivagans*, *Lasiurus cinereus*, *Lasiurus borealis*, *Eptesicus fuscus*, *Tadarida brasiliensis*, *Canis latrans*, *Urocyon cinereoargenteus*, *Procyon lotor*, *Mustela frenata*, *Taxidea taxus*, *Mephitis mephitis*, *Felis concolor*, *Ursus americanus*, and *Odocoileus virginiana*. With the exception of *Eptesicus fuscus*, all bats in this group are migratory taxa. The terrestrial mammals in this group are wide-ranging carnivores or ungulates.

ACKNOWLEDGMENTS

We thank Dr. Robert S. Hoffmann, Museum of Natural History, University of Kansas, Dr. Philip Myers, Museum of Zoology, University of Michigan, and Dr. Victor Diersing, Museum of Natural History, University of Illinois, for allowing us to examine specimens of the San Carlos region under their care. Dr. William R. Anderson, University of Michigan Herbarium, provided valuable assistance in locating field notes and photographs of the 1930 Michigan expedition to the San Carlos Mountains.

The Dirección General de la Fauna Silvestre, under the direction of Sr. Fiacro Martinez, Dr. Antonio Landazuri and Sr. Ignacio Ibarrola, has been of invaluable assistance in providing scientific collecting permits during our several years of study in the region.

It is not feasible to ackowledge, or fully express our appreciation to the nearly 100 undergraduate students who provided much of the groundwork for our study. These students, in groups of 12 to 20, were involved in our department's Vertebrate Field Studies course, WFS 300, which provided the format for an intense faunal study of the area. George Baumgardner provided outstanding assistance on several expeditions with both the basic research project and working with the students. Notable field assistance also was provided by Bill Modi, Ken Wilkins and Duke Rogers. Steve Kelsch assisted with production and editing of the manuscript.

Our deepest gratitude is extended to the hundreds of inhabitants of the region, who befriended us, brought us specimens, showed us new areas, and tolerated our strange ways. Special thanks go to all the people of Rancho San Francisco, especially the brothers Francisco, Maurilio, and Manuel Morales and their families, who provided us food, shelter, introductions and much information about the region. Their friendship made our several stays in the area not only productive, but a rewarding and pleasant cultural experience.

This paper represents contribution number TA-16600 of the Texas Agricultural Experiment Station.

LITERATURE CITED

ALVAREZ, T. 1963. The Recent mammals of Tamaulipas, México. Univ. Kansas Publ., Mus. Nat. Hist., 14:363-473.

ANDERSON, S. 1972. Mammals of Chihuahua: taxonomy and distribution. Bull. Amer. Mus. Nat. Hist., 248:149-410.

BAKER, R. H. 1951. Mammals from Tamaulipas, México. Univ. Kansas Publ., Mus. Nat. Hist., 5:207-218.

———. 1956. Mammals of Coahuila, México. Univ. Kansas Publ., Mus. Nat. Hist., 9:125-335.

———. 1967. A new subspecies of pallid bat (Chiroptera: Verpertilionidae) from northeastern México. Southwestern Nat., 12:329-337.

BARBOUR, R. W., AND W. H. DAVIS. 1969. Bats of America. Univ. Press Kentucky, Lexington, 286 pp.

BAUMGARDNER, G. D., K. T. WILKINS, AND D. J. SCHMIDLY. 1977. Noteworthy additions to the bat fauna of the Méxican states of Tamaulipas (San Carlos Mountains) and Querétaro. Mammalia, 41:237-238.

BIRNEY, E. C. 1973. Systematics of three species of woodrats (genus *Neotoma*) in central North America. Misc. Publ. Mus. Nat. Hist., Univ. Kansas, 58:1-173.

BLAIR, W. F. 1952. Mammals of the Tamaulipan biotic province in Texas. Texas J. Sci., 4:230-250.

BOGAN, M. A. 1975. Geographic variation in *Myotis californicus* in the southwestern United States and México. Fish and Wildlife Serv., Wildlife Res. Rept., 3:1-31.

CHOATE, J. R. 1970. Systematics and zoogeography of Middle American shrews of the genus *Cryptotis*. Univ. Kansas Publ., Mus. Nat. Hist., 19:195-317.

CLARK, W. K. 1953. The baculum in the taxonomy of *Peromyscus boylii* and *P. pectoralis*. J. Mamm., 34:189-192.

COCKRUM, E. L. 1969. Migration in the guano bat, *Tadarida brasiliensis*. Pp. 303-336, *in* Contributions in Mammalogy (J. K. Jones, Jr., ed.). Misc. Publ. Mus. Nat. Hist., Univ. Kansas, 51:1-428.

DALQUEST, W. W. 1953. Mammals of the Méxican state of San Luis Potosí. Louisiana State Univ. Studies, Biol. Sci. Ser., 1:1-229.

DAVIS, W. B. 1974. The mammals of Texas. Bull. Texas Parks and Wildlife Dept., 41 (revised):1-294.

DAVIS, W. B., AND D. C. CARTER. 1962. Review of the genus *Leptonycteris* (Mammalia: Chiroptera). Proc. Biol. Soc. Washington, 75:193-198.

DICE, L. R. 1937. Mammals of the San Carlos Mountains and vicinity. Univ. Michigan Studies Sci. Ser., 12:245-268.

———. 1943. The biotic provinces of North America. Univ. Michigan Press, Ann Arbor, viii+78 pp.

FINDLEY, J. S., AND G. L. TRAUT. 1970. Geographic variation in *Pipistrellus hesperus*. J. Mamm., 51:741-765.

GARDNER, A. L. 1973. The systematics of the genus *Didelphis* (Marsupialia: Didelphidae) in North and Middle America. Spec. Publ. Mus., Texas Tech Univ., 4:1-81.

GENOWAYS, H. H. 1973. Systematics and evolutionary relationships of spiny pocket mice, genus *Liomys*. Spec. Publ. Mus., Texas Tech Univ., 5:1-368.

GENOWAYS, H. H., AND J. K. JONES, JR. 1969. Taxonomic status of certain long-earned bats (genus *Myotis*) from the southwestern United States and México. Southwestern Nat., 14:1-13.

GENOWAYS, H. H., R. J. BAKER, AND J. E. CORNELY. 1979. Mammals of the Guadalupe Mountains National Park, Texas. Pp. 271-322, *in* Biological investigations in the Guadalupe Mountains National Park, Texas (H. H. Genoways and R. J. Baker, ed.), National Park Service, Proc. and Tran. Series, No. 4, 442 pp.

GOLDMAN, E. A. 1943. The races of the ocelot and margay in Middle America. J. Mamm., 24:372-385.

——— . 1950. Raccoons of North and Middle America. N. Amer. Fauna, 60:1-153.

GOODWIN, G. G. 1954. Mammals from México collected by Marian Martin for the American Museum of Natural History. Amer. Mus. Novit., 1689:1-16.

GUZMAN, A. J. 1968. Nuevos registros de murciélagos para Nuevo León, México. An. Inst. Biol. Univ. Nat. Auton. Mexico, 39, Ser. Zool. (1):133-144.

HALL, E. R. 1951a. American weasels. Univ. Kansas Publ., Mus. Nat. Hist., 4:1-466.

——— . 1951b. A synopsis of the North American Lagomorpha. Univ. Kansas Publ., Mus. Nat. Hist., 5:119-202.

——— . 1981. The mammals of North America. 2nd ed. John Wiley and Sons, New York, 1:xviii+1-600+90; 2:viii+601-1181+90.

HALL, E. R., AND J. K. JONES, JR. 1961. North American yellow bats, "*Dasypterus*," and a list of the named kinds of the genus *Lasiurus* Gray. Univ. Kansas Publ., Mus. Nat. Hist., 14:73-98.

HANDLEY, C. O., JR. 1959. A revision of the American bats of the genera *Euderma* and *Plecotus*. Proc. U.S. Nat. Mus., 110:95-246.

HAYWARD, B. J. 1970. The natural history of the cave bat, *Myotis velifer*. Western New Mexico Univ. Res. Sci., 1:1-74.

HOOPER, E. T. 1952. Notes on mice of the species *Peromyscus boylei* and *P. pectoralis*. J. Mamm., 33:371-378.

——— . 1953. Notes on mammals of Tamaulipas, México. Occas. Papers Mus. Zool., Univ. Michagan, 544:1-12.

KOESTNER, E. J. 1941. An annotated list of mammals collected in Nuevo León, México, in 1938. Cont. Zool. Lab. Univ. Illinois, 567:9-15.

KOOPMAN, K. F., AND P. S. MARTIN. 1959. Subfossil mammals from the Gomez Farías region and the tropical gradient of eastern México. J. Mamm., 40:1-12.

LEOPOLD, A. S. 1959. Wildlife of México. The Game birds and mammals. Univ. California Press, Berkley, 568 pp.

LONG, C. A. 1972. Taxonomic revision of the North American badger, *Taxidea taxus*. J. Mamm., 53:725-759.

MARTIN, C. O., AND D. J. SCHMIDLY. 1982. Taxonomic review of the palid bat, *Antrozous pallidus*. Spec. Publ. Mus., Texas Tech Univ., 18:1-48.

MARTIN, P. S. 1958. A biogeography of reptiles and amphibians in the Gomez Farías region, Tamaulipas, México. Misc. Publ. Mus. Zool., Univ. Michigan, 101:1-102.

MARTIN, P. S., C. R. ROBBINS, AND W. B. HEED. 1954. Birds and biogeography of the Sierra de Tamaulipas, an isolated pine-oak habitat. Wilson Bull., 66:38-57.

MODI, W. S. 1978. Morphological discrimination, habitat preferences, and size relationships of *Peromyscus pectoralis* and *Peromyscus boylii* from areas of sympatry in northern México and western Texas. Unpublished M.S. Thesis, Texas A&M Univ., 44 pp.

MÜLLER, C. H. 1939. Relations of the vegetation and climatic types in Nuevo León, México. Amer. Midland Nat., 21:687-729.

MUSSER, G. G. 1968. A systematic study of the Méxican and Guatemalan Gray Squirrel, *Sciurus aureogaster* F. Cuvier (Rodentia: Sciuridae). Misc. Publ. Mus. Zool, Univ. Michigan, 137:1-112.

PACKARD, R. L. 1960. Speciation and evolution of the pygmy mice, genus *Baiomys*. Univ. Kansas Publ. Mus. Nat. Hist., 9:579-670.

SCHMIDLY, D. J. 1972. Geographic variation in the white-ankled mouse, *Peromyscus pectoralis*. Southwestern Nat., 17:113-138.

————. 1973. Geographic variation and taxonomy of *Peromyscus boylii* from México and the southern United States. J. Mamm., 54:111-130.

SCHMIDLY, D. J., F. S. HENDRICKS, AND C. S. LIEB. 1973. Noteworthy additions to the bat fauna of the San Carlos Mountains, Tamaulipas, México. Texas J. Sci., 25:87-88.

SCHWARZ, E., AND H. K. SCHWARZ. 1943. The wild and commensal stocks of the house mouse, *Mus musculus* Linnaeus. J. Mamm., 24:59-72.

SMITH, J. D. 1972. Systematics of the chiropteran family Mormoopidae. Misc. Publ. Mus. Nat. Hist., Univ. Kansas, 56:1-132.

WETZEL, R. M. 1977. The Chacoan peccary, *Catagonus wagneri* (Rusconi). Bull. Carnegie Mus. Nat. Hist., 3:1-36.

WOODBURN, M. O. 1968. The cranial myology and osteology of *Dicotyles tajacu*, the collared peccary, and its bearing on classification. Mem. South. California Acad. Sci., 7:1-48.

WILSON, D. E. 1973. The systematic status of *Perognathus merriami* Allen. Proc. Biol. Soc. Washington, 86:175-192.

VAN GELDER, R. G. 1959. Taxonomic revision of the spotted skunks (Genus *Spilogale*). Bull. Amer. Mus. Nat. Hist., 117:233-392.

VILLA-R., B. 1967. Los murciélagos de México. Inst. de Biól., Univ. Nac. Autónoma México, xvi+491 pp. [For 1966].

YATES, T. L., D. J. SCHMIDLY, AND K. L. CULBERTSON. 1976. Silver-haired bat in México. J. Mamm., 57:205.

APPENDIX 1.—*Mammalian distribution in the San Carlos Mountains according to habitat zones.*

Species	Xeric Thorn Scrub	Thorn Woodland	Deciduous Thickets	Riparian Forest	Pine-oak Forest
Didelphis virginiana	+	+	+	+	
Cryptotis parva				+	
Notiosorex crawfordi				+	+
Mormoops megalophylla				+	
Leptonycteris nivalis				+	
Choeronycteris mexicana				+	
Desmodus rotundus				+	
Myotis velifer	+			+	
Myotis auriculus				+	
Myotis californicus	+		+	+	+
Lasionycteris noctivagans					+
Pipistrellus hesperus	+			+	+
Eptesicus fuscus				+	
Lasiurus borealis	+			+	+
Lasiurus cinereus				+	
Lasiurus intermedius				+	
Lasiurus ega				+	+
Nycticeius humeralis		+		+	+
Plecotus townsendii	+				
Antrozous pallidus				+	+
Tadarida brasiliensis	+			+	
Tadarida macrotis				+	
Dasypus novemcinctus		+	+	+	
Sylvilagus floridanus		+	+	+	+
Sylvilagus audubonii	+				
Lepus californicus	+	+			
Spermophilus mexicanus	+	+			
Sciurus aureogaster				+	+
Sciurus alleni				+	+
Perognathus flavus	+	+			
Perognathus hispidus		+			
Liomys irroratus	+	+	+	+	
Reithrodontomys fulvescens		+		+	+
Peromyscus leucopus		+		+	
Peromyscus pectoralis		+	+	+	+
Peromyscus boylii				+	+
Baiomys taylori		+		+	
Sigmodon hispidus		+		+	
Neotoma micropus	+	+	+		
Canis latrans	+	+	+	+	+
Urocyon cinereoargenteus		+	+	+	+
Bassariscus astutus		+	+	+	+
Procyon lotor		+	+	+	+
Nasua narica		+	+	+	+
Mustela frenata		+		+	
Taxidea taxus	+	+			
Spilogale putorius		+			
Mephitis mephitis	+	+	+	+	

Appendix 1.—*Continued.*

Conepatus leuconotus		+		+	+
Felis concolor	+	+	+	+	+
Felis pardalis	+	+	+	+	+
Felis rufus	+	+	+	+	+
Tayassu tajacu		+	+	+	
Odocoileus virginianus	+	+	+	+	+
Totals	20	30	18	44	24

APPENDIX 2.—*Species of mammals occurng in selected geographic areas of Texas, Nuevo Leon, and Tamaulipas. A plus sign indicates that the species has been recorded from the given area. Records are taken from the literature cited in text and specimens deposited in the Texas Cooperative Wildlife Collections of Texas A&M University.*

Species	San Carlos	Monterrey	Cd. Victoria	Gomez Fárias	Sierra de Tamaulipas	Brownsville	Laredo
Didelphis virginiana	+	+	+	+	+	+	+
Philander opossum			+	+			
Marmosa mexicana				+			
Cryptotis parva	+	+	+	+	+	+	+
Cryptotis mexicana				+			
Scalopus aquaticus						+	
Notiosorex crawfordi	+	+				+	+
Pteronotus parnellii		+	+	+			
Pteronotus davyi		+	+	+			
Mormoops megalophylla	+	+	+		+	+	+
Choeronycteris mexicana	+	+					
Micronycteris megalotis				+			
Carollia perspicillata			+				
Glossophaga soricina			+	+	+		
Leptonycteris nivalis	+	+			+		
Leptonycteris sanborni		+					
Sturnira lilium			+	+	+		
Sturnira ludovici				+			
Artibeus jamaicensis				+			
Artibeus lituratus				+	+		
Artibeus toltecus				+			
Artibeus aztecus				+			
Enchistenes hartii				+			
Centurio senex				+	+		
Desmodus rotundus	+		+	+	+		
Diphylla ecaudata			+	+			
Diaemus youngii				+			
Natalus stramineus			+	+	+		
Myotis auriculus	+	+			+		
Myotis velifer	+	+			+		+
Myotis planiceps		+					
Myotis leibii		+					
Myotis californicus	+	+	+				
Myotis nigricans				+	+		
Lasionycteris noctivagans	+						
Pipstrellus hesperus	+	+	+				
Pipistrellus subflavus				+		+	+
Eptesicus fuscus	+	+	+	+	+		
Lasiurus borealis	+	+	+		+	+	+

APPENDIX 2.—*Continued.*

Species	San Carlos	Monterrey	Cd. Victoria	Gomez Fárias	Sierra de Tamaulipas	Brownsville	Laredo
Lasiurus seminolus		+				+	
Lasiurus cinereus	+	+	+	+		+	+
Lasiurus intermedius	+				+	+	
Lasiurus ega	+	+	+	+	+	+	
Nycticeius humeralis	+	+	+	+	+	+	
Rhogessa tumida			+		+		
Plecotus townsendii	+	+					
Plecotus mexicanus		+					
Antrozous pallidus	+	+		+	+	+	+
Tadarida brasiliensis	+	+	+	+	+	+	+
Tadarida aurispinosa				+			
Tadarida laticaudata				+	+		
Tadarida macrotis	+	+					
Molossus ater			+	+			
Molossus aztecus				+			
Dasypus novemcinctus	+		+	+	+	+	+
Sylvilagus brasiliensis				+			
Sylvilagus audubonii	+	+					+
Sylvilagus floridanus	+	+	+	+	+	+	+
Lepus californicus	+	+	+			+	+
Spermophilus mexicanus	+	+	+			+	+
Spermophilus spilosoma						+	
Spermophilus variegatus		+	+				
Sciurus aureogaster	+		+	+	+		
Sciurus deppei			+	+	+		
Sciurus alleni	+	+	+				
Glaucomys volans				+			
Thomomys bottae		+					
Heterogeomys hispidus				+			
Pappogeomys castanops		+				+	
Perognathus flavus	+	+	+			+	+
Perognathus hispidus	+		+			+	+
Dipodomys compactus						+	
Dipodomys ordii							+
Liomys irroratus	+	+	+	+	+	+	
Castor canadensis						+	+
Oryzomys couesi		+	+	+	+	+	

APPENDIX 2.—*Continued.*

Oryzomys melanotis				+	+		
Oryzomys alfaroi				+			
Oryzomys fulvescens		+	+	+			
Reithrodontomys megalotis		+		+			
Reithrodontomys fulvescens	+	+	+	+	+	+	+
Reithrodontomys mexicanus				+			
Peromyscus maniculatus		+					
Peromyscus leucopus	+	+	+	+	+	+	+
Peromyscus boylii	+	+	+	+	+		
Peromyscus pectoralis	+	+	+	+	+		
Peromyscus eremicus							+
Peromyscus ochraventer				+			
Baiomys taylori	+	+	+	+	+	+	+
Onychomys leucogaster		+				+	+
Sigmodon hispidus	+	+	+	+	+	+	+
Neotoma micropus	+	+	+	+	+	+	+
Neotoma angustapalata				+			
Neotoma albigula		+	+				
Canis latrans	+	+	+			+	+
Canis lupus		+					+
Urocyon cinereoargenteus	+	+	+	+	+	+	+
Bassariscus astutus	+	+	+				
Procyon lotor	+	+	+	+	+	+	+
Nasua narica	+	+	+	+	+	+	+
Potos flavus					+		
Ursus americanus	+	+	+	+			
Eira barbara				+			
Mustela frenata	+	+	+	+	+	+	+
Taxidea taxus	+	+				+	+
Spilogale putorius	+		+		+	+	
Spilogale gracilis		+					+
Mephitis mephitis	+					+	+
Conepatus leuconotus	+		+	+	+	+	+
Conepatus mesoleucus		+					
Felis concolor	+	+	+	+	+	+	+
Felis onca	+	+	+	+	+	+	+
Felis pardalis	+	+	+	+	+	+	+
Felis wiedii				+			
Felis yaguaroundi				+		+	
Lynx rufus	+	+	+			+	+
Tayassu tajacu	+	+	+	+	+	+	+
Odocoileus virginianus	+	+	+	+	+	+	+
Mazama americana						+	

POPULATION ECOLOGY OF SMALL MAMMALS IN THE LLANOS OF VENEZUELA

PETER V. AUGUST

Fleming's studies of small mammal ecology in Central America (summarized in Fleming, 1975) and O'Connell's (1979) research on Venezuelan marsupials constitute the most important long-term investigations of the ecology of small mammal communities in the New World tropics. There are a number of excellent studies of small mammal biology in temperate or arid regions of the Neotropics (Dalby, 1975; Fulk, 1975; Pearson, 1975; Meserve and Glanz, 1978; Pefaur *et al.*, 1979; Mares, 1980; Pizzimenti and De Salle, 1980) that provide valuable data for comparing life history strategies of Neotropical and Nearctic mammals inhabiting equivalent life zones. Similar comparison of temperate *vs.* tropical New World small mammal biology is hampered by the lack of data on species occurring in equatorial habitats. In this chapter, I report the results of two years of small mammal trapping in the Venezuelan llanos. Emphasis is placed on the effects of a heterogeneous habitat and a seasonal environment on small mammal demography and distribution.

MATERIALS AND METHODS

This study was conducted in the llanos of Venezuela on Fundo Pecuario Masaguaral, 45 km. south of Calabozo, Estado Guarico (8°33′ N, 67°37′ W). Masaguaral is over 3000 ha. in size (elevation approximately 75 m.) and is a working cattle ranch. The owner of the ranch, Sr. Tomas Blohm, has maintained Masaguaral as a wildlife sanctuary; fire and hunting pressures have been minimal for over 35 years.

The average annual rainfall of the study area is 1478±66 mm. (mean±SD from 1953 to 1976; data from Corozo Pando, 8 km. south of Masaguaral; data courtesy of the Ministario Ambiente de Recursos Naturales Renovables). Most of the total annual precipitation falls during the wet season, which extends from May to November (Fig. 1). Temperature does not vary appreciably throughout the year; the average monthly maximum temperature is 35.2°C and the average monthly minimum temperature is 21.2°C (Monasterio and Sarmiento, 1976; Troth, 1979).

Study Sites

Masaguaral is a mosaic of different habitats. Important abiotic factors affecting habitat type appear to be soil permeability, soil quality, and the depth of the water table (Sarmiento and Monasterio, 1975; Troth, 1979). Gallery forest occurs along the banks of the Río Guarico (Fig. 2), continuous deciduous forest near the Caño Caracol, and broken deciduous forest (Bajío habitat), seasonally flooded palm-grass savanna (estero habitat), and sandhill woodland (medano habitat) occur throughout the ranch. Troth (1979) provides a thorough description of those habitats on Masaguaral, and general discussions of llanos vegetation are given in Ramia (1967), Sarmiento and Monastario (1969, 1971, 1975), and Monasterio and Sarmiento (1976). Cattle or horses range throughout the ranch.

During the course of this project, I censused small mammals in four distinct habitat types. The following is a brief description of each study site.

Medano Habitat (Dry Grid)

A four-hectare grid was established in medano habitat in September 1976 and was trapped monthly until December 1977. The vegetation of the grid was a mixture of woodland and field (grasses and herbs). Dominant trees and shrubs were *Annona jahnii, Byrsinoma crassifolia, Caesearia hirsuta, Pereskia guamacho, Hymenaea courbaril,* and *Ficus trigonata.* Medano habitat is characterized by having well-drained, sandy soil; however, small flooded depressions were present on the northwest and southeast corners of the grid as well as a narrow strip in the center of the grid. The northwest corner of the grid partially extended into bajío habitat.

Deciduous Forest (Grids DFI and DFII)

Two grids were set in deciduous forest habitat near the Caño Caracol on the east side of Masaguaral. Grid DFI was 2.25 ha. in size and was approximately 1.2 km. from the Caño Caracol. I trapped that grid monthly from January 1977 to April 1978. Grid DFII was 4.0 ha. in size and was adjacent to the Caño. That grid was trapped monthly from January 1978 to May 1978.

Deciduous forest habitat is characterized by a nearly continuous canopy, 7 to 10 m. in height. Vines and thorn-scrub often produced a tangled understory. A complete list of the vegetation of that habitat was given by Troth (1979). Dominant trees of the forest are *Copernicia tectorum, Ficus* spp., *Genipa americana, Pithecellobium carabobense, Guazuma tometosa,* and *Randia venezuelensis.* Wet season flooding is not extensive in deciduous forest habitat.

Palm-grass Savanna (Savanna Grid)

A four-hectare grid was trapped monthly from January 1978 to May 1978 in savanna habitat. The woody vegetation in this portion of Masaguaral (or estero habitat, Troth, 1979) is dominated by *Copernicia tectorum* and thorny shrubs (*Acacia farnesiana*) at the base of many palms. Grasses and sedges (*Panicum laxum, Leersia hexandra* and *Eleocharis* sp.) were the dominant ground cover. This habitat is completely flooded in the wet season.

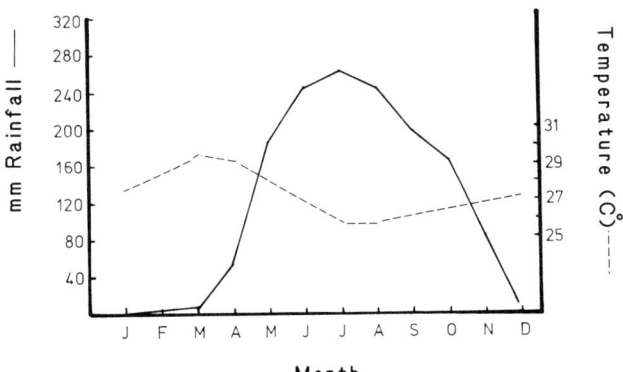

FIG. 1.—Monthly rainfall and average daily temperature for the llanos study area. Rainfall data are monthly means from 1953 to 1976 at Corozo Pando, 8 km. S of the study area. Temperature data are monthly means from 1968 to 1977 at the Llanos Biological Station approximately 40 km. N of the study area. Data provided by the Ministario Ambiente Recursos Naturales Renovables, Caracas.

Bajío (Flooded Grid)

A four-hectare grid was trapped monthly from January 1978 to June 1978 in bajío habitat. Bajíos are mosaics of field and forest patches (matas). A list of the major vegetation types of bajío habitat was given by Troth (1979). Dominant trees and shrubs are *Annona* sp., *Randia venezuelensis*, *Zanthoxylum culantrillo*, *Copernicia tectorum*, *Genipa americana*, *Platymiscium* sp., *Trichilia trifolia*, *Enterolobium cyclocarpum*, and *Pterocarpus acapulcensis*. Lower portions of bajío habitat flooded during the wet season.

Trapping Methods

Each grid consisted of 10 rows and 10 columns of traps spaced at 20-meter intervals except grid DFI, in which traps were spaced at 15-meter intervals.

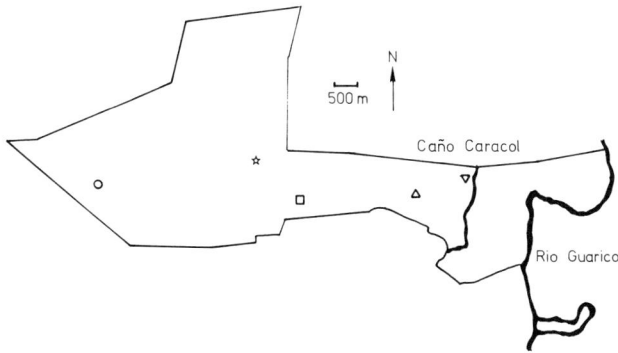

FIG. 2.—Map of Fundo Pecuario Masaguaral showing small mammal study sites. Symbols representing grids are: ○, Savanna; □, Dry; ☆, Flooded; △, DFII; ▽ DFII.

TABLE 1.—*Numbers and types of traps set on the five small mammal study grids.*

Grid	Sherman traps		National traps		Trap nights
	Ground	Tree	Ground	Tree	
Dry	100	20	0	17	15,344
DFI	96	23	0	21	11,200
DFII	100	25	0	24	3,725
Flooded	100	25	0	23	4,400
Savanna	100	10	12	0	3,660

Each grid contained 100 ground trap stations (Table 1), except grid DFI which had 96 (4 stations were eliminated from the northeast corner of the grid). At each trap station, one Sherman live trap was placed on the ground. Systematically placed throughout each grid were Sherman live traps and squirrel-size National live traps wired to branches or lianas at approximately 2-3 m. above the ground.

Grids DFI, DFII, Savanna, and Flooded were censused five consecutive days monthly and the Dry grid was trapped seven days each month. In the wet season, Sherman tree traps were baited with a mixture of corn and plátano or banana, National traps were baited with plátano or banana alone. Sherman ground traps were baited with corn. In the dry season, corn was replaced with dry dog food pellets. Ant disturbance was noticeably less in traps baited with dog food and this did not seem to affect capture frequency of small mammals in this region (J. Gomez-Nunez, personal communication). One drawback of using dog food was that when it became wet, it would disintegrate into a loose mush that would often jam the treadle mechanism of the traps, hence it was used only in the dry season.

Traps were checked in the morning between 0700 and 1000 hours. Data taken on each captured animal were: species, toe-clip number, sex, reproductive condition, age, weight, length of hind foot, and general physical condition. Reproductive conditions recorded were: males—testes abdominal, testes scrotal; females—vulva closed, vulva perforate, pregnant, or lactating. All captured animals were released at the point of capture immediately after examination. The behavior of animals after being released was usually recorded.

Habitat Selection

Habitat variables were measured at each trap station (Table 2). When possible, quantitative measurements were made, but for some aspects of habitat physiognomy this was not possible. In recording foliage density in different vertical strata (CAN, MID, SHRUB), I used a qualitative scaling system, with zero indicating no foliage and four indicating dense vegetation. Because I was the only person recording habitat variables in the study, any biases were consistent across grids. The canopy stratum is defined as the uppermost layer of vegetation; the shrub layer refers to the stratum occupied

TABLE 2.—*Habitat variables measured at each grid station (Mnemonic abbreviations in parentheses).*

Habitat variable	Description
Canopy height (CANHT)	Canopy height (in meters) above trap station
Canopy density (CAN)	Density of canopy on a scale from 0, sparse canopy vegation, to 4, dense canopy vegetation
Midstory density (MID)	Density of vegetation in midstory stratum; scaling same as CAN
Shrub density (SHRUB)	Density of shrub layer vegetation; scaling same as CAN
Percentage ground cover (GC)	Percentage of herbaceous ground cover in a 3.5-meter radius around trap station
Ground cover height (GCHT)	Average height (in centimeters) of herbaceous vegetation in a 3.5-meter radius around trap station
Diameter at breast height (DBH)	Diameter (in centimeters) of all trees and shrubs (with DBH greater than 3.2 cm) within a 3.5-meter radius of trap
Number in DBH sample (NDBH)	Number of trees and shrubs in DBH sample
Number of species in DBH sample (NSPP)	Number of tree and shrub species in DBH sample
Percentage canopy cover (PCCAN)	Percentage canopy cover above trap station.

by low shrubs and vines; the midstory is the vegetation layer between the canopy and shrub strata. For all statistical analyses, the variables recorded as percentages (GC and PCCAN) were arcsine transformed (Sokal and Rohlf, 1969) and these transformed variables are referred to as TGC and TPCCAN, respectively. Canopy height was estimated on all grids. However, on the Flooded grid, heights were measured using a Brunton pocket transit. Estimated canopy heights were highly correlated with measured canopy heights (Pearson $r=0.74$, $P<0.001$, $N=48$). The mean ($\pm$SD) difference between estimated and measured canopy heights was 1.25 ± 2.5 m. Correlation of the difference between estimated and measured canopy height was nonsignificant (Pearson $r=0.11$, $P>0.05$) indicating no relationship between canopy height and error in estimating height. Habitat variables were measured in the dry season for all grids, and both dry and wet season data were taken on grids Dry and DFI. For analysis of microhabitat selection and partitioning, the original habitat measures were reduced to a set of relatively nonredundant variables. These were canopy height (CANHT), canopy density (CAN), midstory density (MID), shrub density (SHRUB), ground cover height (GCHT), diameter at breast height (DBH), number of trees and shrubs in DBH sample (NDBH), and the arcsine transformation of the percentage ground cover (TGC).

At each tree trap, a number of habitat variables were recorded: TRAPHT, the height of the trap in cm.; TRAPSUP, the type of vegetation that supported the trap (for example, tree, shrub, liana); TRAPON, the part of the plant that supported the trap (for example, branch, trunk, liana); BRANCH2, where the supporting vegetation ultimately led (for example, tree canopy, ground, space); ANGLE, estimated trap angle in degrees; DEDALV, whether the vegetation supporting the trap was dead or alive; TOPALM, whether the vegetation supporting the trap led to a palm canopy or not; DPALM, the distance to the closest palm; DFIG, the distance to the nearest *Ficus* sp; and CANHT, the canopy height. The variables DFIG and DPALM were reduced to four distance classes for analyses (in meters): 0-2, class 1; 3-9, class 2; 10-19, class 3; >20, class 4.

Habitat selection was tested using discriminant analysis. My null hypothesis was that there was no difference between small mammal microhabitat and available habitat. I quantified available habitat by using the mean of each habitat variable for all trap stations regardless of whether an animal was captured or not. Univariate analysis of variance (ANOVA) was used to test the equivalence of available habitat and habitat profiles of each species. The significance of Wilks λ for the discriminant analysis was used to test this null hypothesis with all variables considered simultaneously. This is equivalent to a multivariate analysis of variance (Neff and Marcus, 1980). Box's test of the homogeneity of group variance-covariance matrices was performed prior to each discriminant analysis (Cooley and Lohnes, 1971). Generally, the results indicated that the matrices were not homogeneous and thus a basic assumption of discriminant analysis was not met. However, this seems to be the rule rather than the exception with ecological data (Green, 1971, 1980; M'Closkey, 1976; Dueser and Shugart, 1978). Discriminant analysis is a robust test and performs reliably even when data are not multivariate normal or have heterogeneous variance-covariance matrices (Lachenbruch, 1975; Neff and Smith, 1979).

To determine if arboreal or scansorial species were selecting certain features of the tree or shrub habitat, I tested the null hypothesis that there was no difference between species habitat profiles and available habitat. For this analyis I used one-way ANOVA to test the equivalence of group means of continuous habitat variables and χ^2 tests of homogeneity of two-way contingency tables (species X habitat variable states). As the sample size of most of the χ^2 analyses was less than 100, Yates correction factor was used (Sokal and Rohlf, 1969). When presented in the text, means are followed by ± one standard deviation. Computations were done using programs in the Statistical Analysis System (SAS, Barr *et al.*, 1979) and the Statistical Package for the Social Sciences (SPSS, Nie *et al.*, 1975).

RESULTS

Trap Response

In 38,329 trap nights, 197 individuals of nine species of small mammals were captured 387 times. This represents a capture success of approximately one per cent. The number of each species captured per grid, sex ratios, distribution of captures over the different types of traps used, and average number of captures per individual are presented in Table 3.

TABLE 3.—*Summary of capture results for llanos small mammals. Sex ratios are expressed as the ratio of males : females. Abbreviations are: S, Sherman trap; N, National trap; Caps, captures.*

Grid/species	Males	Females	Sex ratio	Total individs	Grnd S traps	Tree S traps	Tree N traps	Total caps	Caps/ individual
Dry									
Didelphis	10	7	1.4	17	3	0	36	39	2.35
Marmosa	11	3	3.7	14	3	27	3	33	2.36
Zygodontomys	8	8	1.0	16	29	0	0	29	1.82
Oryzomys	1	0		1	0	1	0	1	1.00
Sigmomys	1	0		1	1	0	0	1	1.00
Echimys	1	0		1	0	0	2	2	2.00
DFI									
Didelphis	1	0		1	0	0	2	2	2.00
Marmosa	20	20	1.0	40	18	68	5	91	2.25
Zygodontomys	5	7	0.7	12	29	0	0	29	2.41
Heteromys	11	11	1.0	22	39	0	0	39	1.77
Echimys	2	1	2.0	3	0	2	1	3	1.00
Sciurus	3	2	1.5	5	0	0	7	7	1.40
DFII									
Marmosa	3	4	0.75	7	2	9	0	11	1.57
Zygodontomys	0	1		1	2	0	0	2	2.00
Heteromys	4	2	2.0	6	6	0	0	6	1.00
Sciurus	1	0		1	0	0	1	1	1.00
Flooded									
Marmosa	9	4	2.25	13	4	37	0	41	3.15
Zygodontomys	9	12	0.75	21	24	0	0	24	1.38
Sigmomys	0	2		2	2	0	0	2	1.00
Rhipidomys	2	1	2.0	3	0	4	0	4	1.33
Savanna									
Oryzomys	6	4	1.5	10	1	19	0	20	2.00

Sex ratios of most species were equal; however, there were some notable exceptions (Table 3). There were nearly four times as many males of *Marmosa robinsoni* captured on the Dry grid as there were females ($\chi^2 = 4.6$, $P<0.05$). The sex ratio of *Marmosa* on the Flooded grid was 2.25 in favor of males. The overall sex ratio of *Marmosa* (grids pooled) was 1.39 and this did not differ significantly from unity ($\chi^2 = 1.9$, $P>0.05$).

The distribution of captures between tree and ground traps provides a measure of the arboreal/terrestrial behavior of each species (Table 3). Ninety-two per cent of all *Didelphis marsupialis* captures were in National traps set in trees, the remaining captures were in ground-set Sherman traps. Eighty-four per cent of all *Marmosa robinsoni* captures were in tree traps and primarily in tree-set Sherman traps (81.5 per cent). *Zygodontomys brevicauda*, *Heteromys anomalus*, and *Sigmomys alstoni* were captured exclusively in ground set Sherman traps. Nearly all captures of *Oryzomys bicolor* were in tree-set Sherman's (95.3 per cent), and all captures of

TABLE 4.—*Monthly MNKA densities (in number per hectare) of small mammals on the Dry grid. Species abbreviations are: D. m.* Didelphis marsupialis; *M.r.* Marmosa robinsoni, *S. a.* Sigmomys alstoni, *Z. b.* Zygodontomys brevicauda, *O. b.* Oryzomys bicolor, *H. a.* Heteromys anomalus, *R. sp.* Rhipidomys sp., *S. g.* Sciurus granatensis.

Month	D. m.	E. s.	S. a.	M. r.	Z. b.	O. b.
1976						
Sep	2.50	0	0	0	1.00	0
Oct	1.50	0	0	0.50	0.25	0
Nov	0.75	0	0	0.25	0.25	0
Dec	1.25	0	0	0.25	0	0
Jan	0.75	0	0	1.00	0	0
Feb	0.25	0.25	0	0.75	0.25	0
Mar	0.50	0.25	0	1.25	0	0
Apr	0	0.25	0	0.75	0	0
May	0	0.25	0	1.00	0	0
Jun	0	0.25	0	0.50	1.75	0
Jul	0	0.25	0	0.25	1.00	0
Aug	0	0	0.25	0.25	0.50	0
Sep	0	0	0	0.25	0.25	0
Oct	0.25	0	0	0.50	0	0.25
Nov	0	0	0	0.25	0	0
Dec	0	0	0	0.25	0	0

Echimys semivillosus, Sciurus granatensis, and *Rhipidomys* sp. were in tree traps. Recapture rates were highest for the marsupials, *Marmosa* and *Didelphis. Zygodontomys* had a high rate of recapture (2.4 captures/individual) on grid DFI. All other species were generally captured less than twice per individual.

TABLE 5.—*Small mammal densities (based on MNKA) on grid DFI. Abbreviations given in Table 4.*

Month	D. m.	E. s.	S. g.	H. a.	M. r.	Z. b.
1977						
Jan	0.44	0	0	0.89	4.00	0.89
Feb	0.44	0.44	0	0.89	5.78	0.89
Mar	0	0	0	0.44	4.44	0.89
Apr	0	0	0.44	0.44	3.56	0.89
May	0	0.44	0.44	0.44	1.33	1.33
Jun	0	0	0.89	0.89	0.89	1.78
Jul	0	0.44	0.44	1.33	0.44	0
Aug	0	0	0.44	0	0.89	1.33
Sep	0	0	0.44	1.33	2.22	0.44
Oct	0	0	0.44	1.78	2.67	0.44
Nov	0	0	0.44	0	2.67	0.44
Dec	0	0	0.44	1.33	3.11	0
Jan	0	0	0.44	1.33	3.56	0
Feb	0	0	0.44	2.22	3.11	0
Mar	0	0	0.44	1.78	4.44	0
Apr	0	0	0.44	2.22	0.89	0

Density

To extrapolate density estimates from live-trapping data, it must be shown that: 1) all animals have an equal probability of capture, 2) there is no differential loss or gain of individuals in the population during the census period, and 3) the total area sampled by traps is known. Considering the low capture success (1 per cent), there was an abundance of traps available to an individual rodent or opossum. However, trap disturbance by ants, termites, rain, etc. resulted in fewer traps available to small mammals than the sampling effort (trap nights) indicates. Without knowing the true abundance of small mammals on the grids, it is impossible for me accurately to assess age or sex differences in trap response (see Mares *et al.*, 1981, for a discussion of this important problem). Certainly, individuals were born, died, and moved out of or onto grids during the course of the study. However, these changes were likely small within each day of the 5-day census periods. The most serious violation of these assumptions regards knowledge of the total area sampled by traps. Considering the arboreal nature of many of the species, a total volume sampled may be more appropriate. August (1981) discussed in detail the methods available to estimate accurately sampling area.

The densities reported here are based on a grid size that includes a boundary strip width equal to half of the intertrap distance. Although the density values might overestimate true density, they are comparable to the results of other small mammal trapping studies. The monthly minimum number known alive (MNKA) for each species on each grid is presented in Tables 4 to 7. *Didelphis* was moderately abundant on the Dry grid at the beginning of the study but after March 1977 was caught only once. *Echimys*, *Sigmomys*, and *Oryzomys* were never common on the Dry grid. *Marmosa* was a moderately common species in this habitat. *Zygodontomys* was frequently captured, especially at the beginning of the wet season.

Didelphis was infrequently captured on grid DFI. The trapping data indicate that *Echimys* was present only periodically on grid DFI. This is an underestimate of true *Echimys* abundance. One female nesting in a hollow tree branch approximately 2.5 m. from the ground was known to be present on the grid from February 1977 to approximately March 1978. She was captured once in February 1977 but was not recaptured for the rest of the study. *Heteromys* was captured regularly throughout the study. *Marmosa* was the most abundant small mammal on the grid and reached densities of over five individuals per hectare. *Zygodontomys* was captured in low numbers throughout most of the monthly censuses on the DFI grid; however, it was not captured in the final five months of trapping. *Sciurus* was regularly captured, albeit in low number throughout the study.

Grid DFII was sparsely populated by *Marmosa*, *Heteromys*, *Zygodontomys*, and *Sciurus*. None of these taxa was abundant, and only one or two individuals were captured in monthly censuses.

TABLE 6.—*Small mammal densities (based on MNKA) for the Flooded grid. Abbreviations given in Table 4.*

Month	M. r.	R. sp.	S. a.	Z.b.
1978				
Jan	2.00	0.25	0	0.50
Feb	2.25	0	0	0
Mar	1.75	0	0	0.25
Apr	1.25	0	0	0
May	1.25	0.25	0	0.50
Jun	0.25	0.25	0.50	2.75

The Flooded grid maintained a moderate population of *Marmosa*. *Zygodontomys* was rare until the beginning of the wet season, when capture rates markedly increased. *Rhipidomys* and *Sigmomys* were captured infrequently.

Oryzomys was the only species captured on the Savanna grid. Density ranged from 0.25 mice/ha. to 1.25 mice/ha. over the dry season of 1978. The Savanna grid had the lowest average monthly small mammal density (0.7 individuals/ha.). The mean monthly densities for the other grids (species pooled, in individuals/ha.) were: DFI, 4.9; Flooded, 2.3; Dry, 1.4; and DFII, 0.9.

Small mammal abundance varied seasonally. There were sufficient captures of *Zygodontomys* and *Marmosa* to examine seasonal trends in capture frequency, and these are presented in Fig. 3. Captures of *Zygodontomys* dramatically increased with the onset of the rainy season in 1977 and 1978. Captures declined through the wet season and remained low until the beginning of the following wet season. *Marmosa* capture rates showed an opposite pattern and were highest in the dry season, then declined during the initial portion of the wet season.

Reproduction, Age Structure, and Recruitment

Given the small monthly sample size of most of the mammal species captured in this study, it is difficult to make definitive statements on

TABLE 7.—*Small mammal densities (based on MNKA) for 1978 on grids DFII and Savanna. Abbreviations given in Table 4.*

Month	DFII				Savanna
	H. a.	M. r.	S. g.	Z. b.	O. b.
Jan	0	0.25	0	0	1.00
Feb.	0	0.75	0.25	0	0.50
Mar	0.25	0.50	0	0	1.25
Apr	0.50	0.50	0	0	0.75
May	0.50	0.50	0	0.50	0.25
Jun					0.25

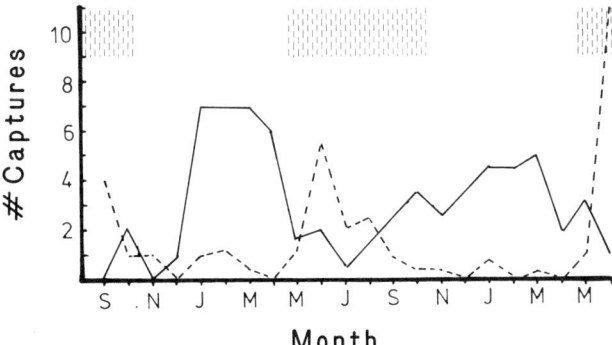

Fɪɢ. 3.—Monthly capture rates of *Zygodontomys brevicauda* (broken line) and *Marmosa robinsoni* (solid line). The stipling above the graph indicates the duration of the wet season.

breeding patterns. However, concurrent examination of age structure and the presence of breeding individuals in the population permits a general analysis of reproductive timing in three species that were captured regularly, *Marmosa robinsoni*, *Zygodontomys brevicauda*, and *Heteromys anomalus*.

Two age classes were recognized, subadults and adults. Aging criteria included foot size, body mass, and pelage characters. *Marmosa* were sexually dimorphic; adult males ($\bar{X}$ weight, 62.9±17.9 g., N=95) were larger than adult females ($\bar{X}$ weight, 42.3±8.5 g., N=32; one-way ANOVA, F=39.3, P<0.001).

Marmosa.—Because male *Marmosa* have permanently descended testes, it was difficult to judge breeding condition in the field. Males did have a conspicuous sternal gland (Barnes, 1977), which was active only during part of the year. That gland was most prominent toward the end of the dry season. The seasonal activity of the sternal gland suggests that it has some function in reproduction or social behavior. Pregnant or lactating female *Marmosa* were rarely captured. Pregnant females, or females with attached young, were captured in April 1977 (4 individuals), May 1978 (2 individuals), May 1978 (2 individuals), and June 1977 (1 individual). Lactating females without attached young were captured in May 1977 (1), June 1977 (1), and June 1978 (1). Table 8 gives litter sizes and crown-rump lengths of neonates attached to females. Subadult *Marmosa* appeared on the grids in the beginning of the wet season (Fig. 4). From these data, it is apparent that breeding begins at the end of the dry season and young are weaned at the beginning of the wet season.

Zygodontomys.—Breeding *Zygodontomys* were taken throughout the year. Fig. 5 presents the months in which breeding adult males and females were captured and the monthly proportion of individuals in the subadult age class. One pregnant female taken on 25 March 1977 contained 4 embryos with crown-rump lengths of 3 mm. On 8 January, I discovered a nest of five neonates under a hummock of grass on the Flooded grid. At the time of

TABLE 8.—*Litter size and neonate crown-rump length (in millimeters) for* Marmosa robinsoni. *The dates indicate when each female with attached young was captured. Crown-rump lengths are the average of each litter.*

Date	Number of neonates	Crown-rump length
9 June 1977	15	10
23 April 1977	16	6
8 May 1978	15	7
14 May 1978	3	14

discovery (approximately 1000 hours), the young were lying atop a bed of cut grass in a depression in the hardened soil (not in a burrow). Of the five, three were female and two were male; each was covered by short, sparse, gray pelage.

Heteromys.—Male *Heteromys* with scrotal testes were captured in nearly all months. Pregnant females were captured in June 1977 (two individuals), September 1977 (1), December 1977 (1), and May 1978 (1). Lactating females were captured in October 1977 (2) and February 1977 (1). Most subadult *Heteromys* appeared in the traps in the middle of each dry season (Fig. 6).

Persistence on Grids

My estimate of persistence is the interval (in days) between the first capture of an individual and its last capture. Using unfenced grids, it is impossible to ascertain whether an individual ceased being captured because it died, moved off the grid, or stopped entering traps. The mean number of days that each species remained on each grid is summarized in Table 9.

There was considerable variation in persistence between grids for most species, but no differences were statistically significant (Kruskal-Wallis test for multigroup cases, Wilcoxon two-sample test for two group cases). To test if persistence on grids was equal between adults and subadults, I compared, by species, average persistence of individuals that were subadult at first capture to persistence for individuals that were adult at first capture. Similarly, I tested for equivalence of persistence between individuals that were subadult at the time of last capture with persistence of those that were adults at last capture. In both analyses, no significant differences were found for any species. There was no difference in persistence between sexes in any of the rodents or opossums captured.

I examined seasonal effects by comparing the mean persistence of individuals for which first capture was in dry season against those for which first capture was in the wet season. Similarly, I compared persistence estimates of individuals for which the last capture was in the wet season against those for which the last capture was in the dry season. In both analyses, wet season persistence equalled dry season persistence for all species.

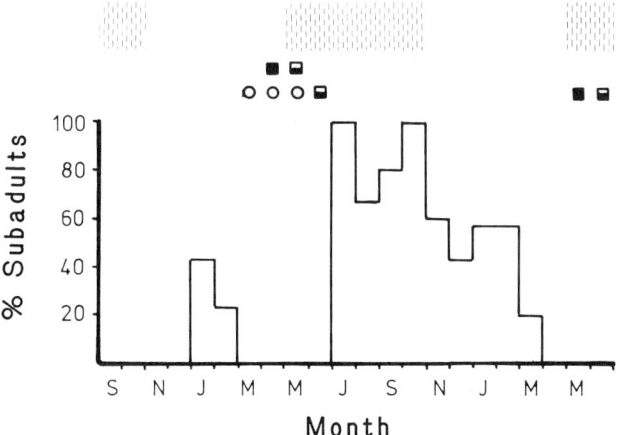

FIG. 4.—Reproduction and monthly age structure for *Marmosa robinsoni*. The histogram represents the proportion of subadults in a monthly census. Months in which reproductively active females (open squares), pregnant females (closed squares), lactating females (half-closed squares), and males with sternal glands (open circles) were captured are indicated. Open circles represent scrotal males in the other species.

I predicted that persistence should be negatively correlated with age; young animals would remain on a grid longer than older individuals. In this anaysis, I used foot size as a measure of age. There was a marginally significant negative correlation (Spearman $r=-0.38$, $N=22$, $P<0.05$) between persistence and foot size at first capture for *Heteromys*. There were no significant correlations between age and persistence for the other species.

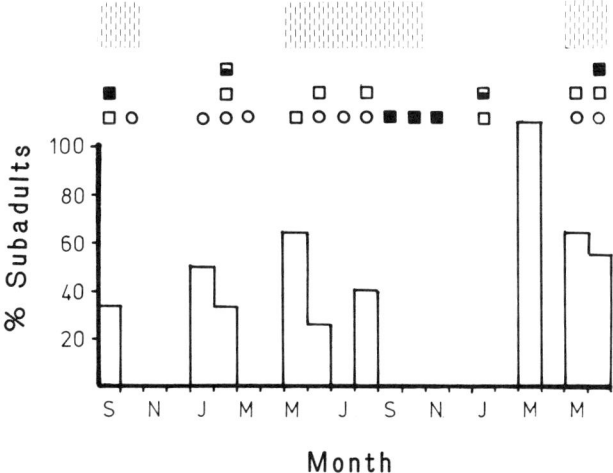

FIG. 5.—Reproduction and age structure in *Zygodontomys brevicauda*. See Fig. 4 for explanation of symbols.

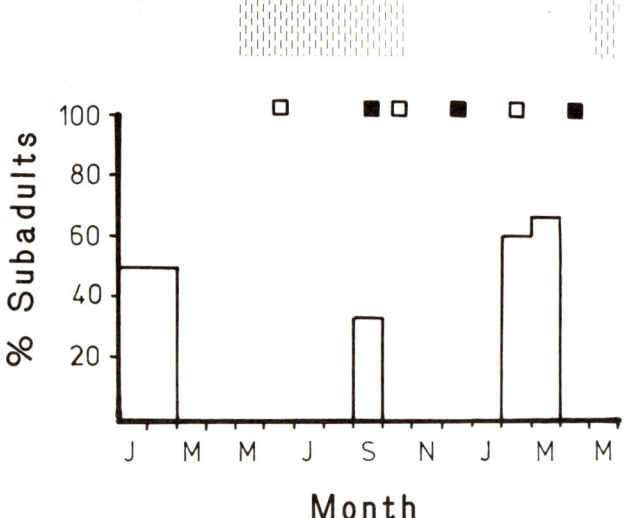

FIG. 6.—Reproduction and age structure for *Heteromys anomalus*. See Fig. 4 for explanation of symbols.

Movements, Home Range, and Spacing Behavior

My estimate of movements is based on the linear distance moved between successive captures for each individual. Using the Wilcoxon two-sample test for two-group comparisons and the Kruskal-Wallis test for multiple-group comparisons, I compared movements between sexes, study grids, ages, seasons, reproductive conditions, and time between captures for species with more than 15 captures. Mean distances moved for each species on each grid is summarized in Table 10. There were significant differences in movements between grids for *Marmosa* ($P<0.05$) and *Zygodontomys* ($P<0.01$). It is important to note that linear distance moved by arboreal or scansorial species such as *Didelphis* and *Marmosa* is somewhat artificial inasmuch as those taxa occur in a habitat volume rather than a two dimensional plane and their actual movements contain a vertical dimension.

For all species, there was no significant difference in movements made by males and females; however, males were slightly more vagile than were

TABLE 9.—*Persistence data for llanos small mammals. Data presented are mean±SD days spent on a grid. Maximum values are given in parentheses followed by sample size.*

Species	Dry	DFI	DFII	Flooded	Savanna
Didelphis	25 ± 35 (103) 16				
Marmosa	37 ± 88 (328) 14	41 ± 49 (175) 40	13 ± 15 (39) 7	49 ± 53 (146) 13	
Zygodontomys	8 ± 14 (35) 16	24 ± 37 (91) 12		1 ± 1 (4) 21	
Heteromys		23 ± 34 (129) 22	0 ± 0 (0) 6		
Sciurus		54 ± 75 (146) 5			
Oryzomys					16 ± 36 (89) 11

TABLE 10.—*Average distance moved (in meters) between successive captures for llanos small mammals. The standard deviation is given followed by the sample size in parentheses. "Pooled" represents species averages over all grids.*

Species	Dry	DFI	DFII	Flooded	Savanna	Pooled
Didelphis	59 ± 25 (22)	77 ± 0 (1)				61 ± 25 (23)
Marmosa	69 ± 48 (19)	39 ± 29 (50)	67 ± 67 (4)	43 ± 28 (28)		47 ± 36 (101)
Zygodontomys	23 ± 22 (13)	47 ± 24 (17)	109 ± 0 (1)	16 ± 19 (8)		34 ± 28 (39)
Heteromys		32 ± 28 (18)				32 ± 28 (18)
Oryzomys					6 ± 14 (11)	6 ± 14 (11)
Sciurus		66 ± 14 (2)				66 ± 14 (2)
Echimys	29 ± 0 (1)					29 ± 0 (1)

females. There were no statistical differences in adult versus subadult movements for any species, although adults were generally more motile. Distances moved in the dry season were equal to those in the wet season.

To test if breeding condition affected vagility, I compared average distance moved by nonbreeding females, females with attached young, and lactating females of *Marmosa* and between nonbreeding, pregnant, and lactating females of *Zygodontomys*. In neither species was there a significant difference in distance moved among reproductive classes. However, in both species, breeding females consistently moved greater distances between captures than nonbreeding females.

I predicted that the distance moved between captures should be related to the amount of time that had elapsed since an individual's last capture. To test for this, I calculated the average distance moved when successive captures were in the same census period (month), average distance moved between captures in successive census periods, average distance moved when two months had elapsed since last capture, *etc.* There was a significant difference in movements among time intervals from last capture for *Zygodontomys* ($P<0.05$), and, as predicted, movements increased as the time interval between captures increased. There was no difference in movements among the time interval categories for the other species examined. This rejection of my initial hypothesis could be an indication that the other taxa are rather site specific and maintain more or less permanent home range areas.

I predicted a positive relationship between vagility and body size; simply, larger species should move farther between captures than do smaller species. The data show a positive correlation between body size and average distance moved when all grids are pooled (Fig. 7).

A number of elegant models are available for calculating home range size in small mammals, but they generally require a moderately large number of captures per individual, a luxury not available in the present study. Nevertheless, simple estimates of home range can be determined, and variation in ranging patterns can be examined. I calculated home range size by plotting capture points for all individuals with more than three recaptures on grid maps, connected the outer ones so that a minimum area

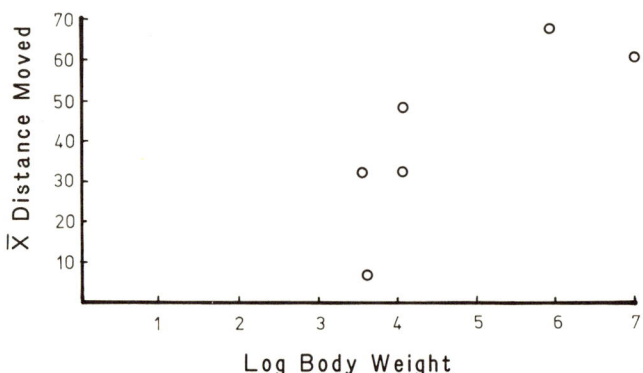

FIG. 7.—Relationship between body size and the average distance moved between successive captures for small mammals.

polygon was formed, and determined its area with a K and E polar planimeter. Nonparametric statistics were used in the analysis of home range data. The Wilcoxon two-sample test was used in all two-group tests, and the Kruskal-Wallis test was used in all multigroup tests.

Home range estimates for each species for which there were enough captures to create a polygon are presented in Table 11. Again, the disclaimer must be made that arboreal or scansorial species such as *Didelphis* and *Marmosa* actually occupy home range volumes rather than home range planes. Home ranges for those species will be reported as planes (in square meters, m^2) because it was impossible to estimate from trap data alone the extent to which they move about vertically in the vegetation.

Marmosa home ranges were largest on the Dry grid and smallest on grid DFI, but this difference was not statistically significant ($P>0.05$). Home ranges of females ($\bar{X}=1970$ m^2) were larger than those of males (974 m^2) but not significantly so ($P>0.05$). Home ranges of *Marmosa* were greater in the dry season ($\bar{X}=1515$ m^2) than they were in the wet season (993 m^2), but again, the difference was not statistically significant ($P>0.05$). Home range size for *Marmosa* was correlated significantly with foot size at first capture (a measure of age; Spearman $r=0.63$, $P<0.01$) and the number of captures per individual (Spearman $r=0.67$, $P<0.001$).

TABLE 11.—*Home range data for llanos small mammals. Mean home ranges (in square meters) are given followed by one standard deviation and the sample size in parentheses.*

Species	Dry	DFI	Flooded	Savanna
Marmosa	3596 ± 5035 (3)	1286 ± 1224 (13)	1727 ± 490 (4)	
Didelphis	1991 ± 1231 (6)			
Zygodontomys	400 ± 0 (1)	1317 ± 354 (4)		
Heteromys		182 ± 120 (3)		
Oryzomys				410 ± 0 (1)

There was a significant difference in home range size among species ($P<0.05$); means (sample size in parentheses, grids pooled) and standard deviations for each species are: *Didelphis*, 2007±1236 m² (6); *Marmosa*, 1721±2082 m² (20); *Zygodontomys*, 1134±512 m² (5); and *Heteromys*, 181±119 m² (3). The scansorial species had larger home ranges than did the terrestrial species; however, this could be due to size factors—at least for *Didelphis* (McNab, 1963). Home range size for *Zygodontomys* could have been biased by grid geometry. There was a significant correlation between home range area for that species and the length of the home range that ran along the edge of the grid (Spearman $r=-0.89$, $P<0.05$). Similar relationships between home range size and grid geometry were not evident for the other species examined.

For the analysis of spacing behavior, I superimposed home range polygons of all individuals with more than three recaptures onto grid maps. There were sufficient captures of *Marmosa* for the Flooded grid in the dry season of 1978 and grid DFI for the dry season of 1977, wet season of 1977, and dry season of 1978 to carry out this analysis but insufficient data for the other species. Home range overlap was quantified by creating a data matrix, the cells of which (O_{ij}) are the per cent of the home range of individual j that is overlapped by individual i. The mean of all cells in the matrix is reported as the mean home range overlap for a given species in a given season.

It should be noted that analysis of spacing behavior based on multiple live trap captures contains a number of biases. First, the patterns of spatial use are based on a small number of observation points (capture localities). Second, each observation point represents a capture site and does not indicate relative use of an area. All capture points are weighted equally, and I cannot distinguish between regular movement paths and unsual forays beyond the normal range of an animal. Third, temporal partitioning of space use is not indicated in using capture points. Fourth, only those animals with sufficient number of recaptures were used in the analysis. It is possible that an individual's home range contained other individuals that escaped capture or were not captured frequently enough to be considered in the analysis. Those biases might interact to show home range overlap where in fact little occurs. Conversely, if ranges regularly abut one another, this could be taken as preliminary evidence of mutually exclusive home ranges.

There was considerable overlap of *Marmosa* home ranges on grid DFI in the dry season of 1977 (Fig. 8A). The mean range overlap was 15.4±17.2 per cent. Home ranges of males overlapped those of other males by 13.7±15.4 per cent, and ranges of females overlapped those of other females by 17.9±20.5 per cent. There was least home range overlap (3.2±5.7 per cent) among individual *Marmosa* on grid DFI during the wet season of 1977 (Fig. 8B). Total home range overlap was 3.1±5.3 per cent among *Marmosa* on grid DFI in the dry season of 1978 (Fig. 8C), and overlap among males was

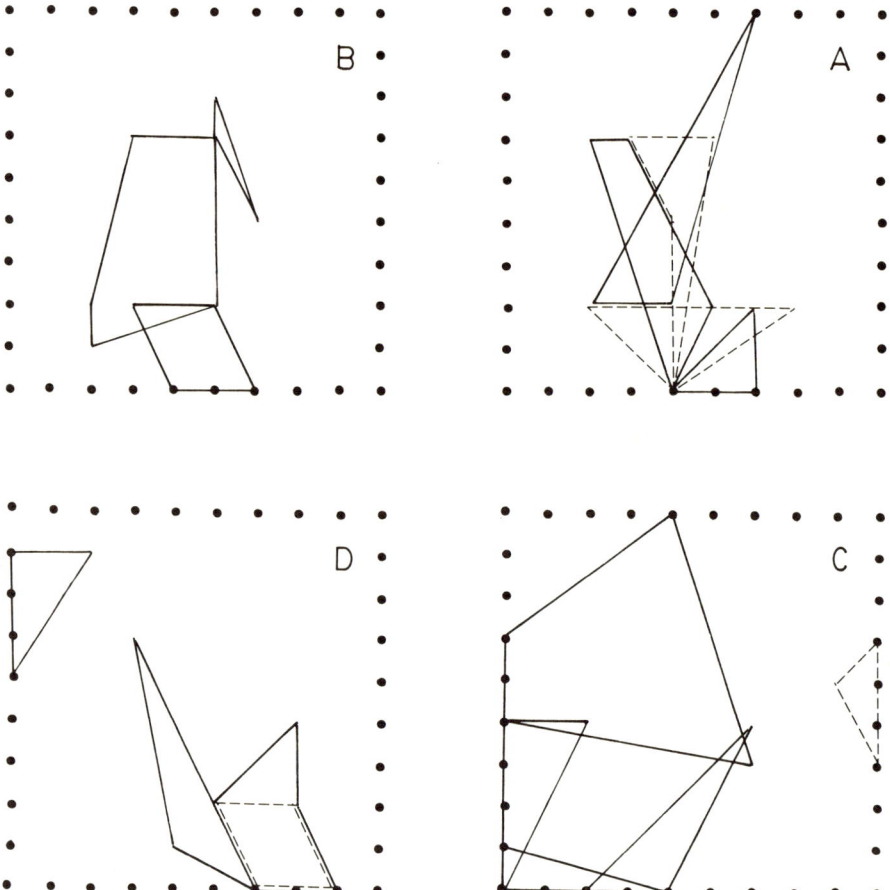

Fig. 8.—Home ranges of *Marmosa robinsoni* on: **A**, grid DFI during the dry season 1977; **B**, grid DFI during the wet season 1977; **C**, grid DFI during the dry season 1978; **D**, the Flooded grid during the dry season 1978. The dots indicate the position of the outermost rows and columns of traps. Polygons represent the home range of one individual. Males are indicated by solid lines; females, by dotted lines.

4.0±6.2 per cent during that period. A similar pattern of exclusive home ranges was evident among *Marmosa* on the Flooded grid during the dry season of 1978 (Fig. 8D). Total overlap was 13.9±33.3 per cent for all individuals and 7.4±22.3 per cent among males.

Habitat Selection

The five grids were different in overall habitat profile (Table 12). The Dry grid was a mosaic of grass fields and forest patches. Canopy height was low and variable, and the ground and midstory vegetation strata were poorly developed. Herbaceous ground cover was dense and tall. Grids DFI and DFII were similar in physiognomy; both had tall canopies and dense

TABLE 12.—*Mean (standard deviations in parentheses) values for the habitat variables for each grid. Seasons are indicated beneath the grid name.*

Variable	Dry		DFI		DFII	Flooded	Savanna
	Dry	Wet	Dry	Wet	Dry	Dry	Dry
CANHT	1.7 (2.8)	1.7 (2.8)	10.2 (4.1)	10.2 (4.1)	8.5 (3.2)	3.4 (4.4)	0.0 (0)
CAN	0.4 (0.6)	0.4 (0.8)	1.1 (0.5)	1.4 (0.6)	1.4 (0.7)	0.6 (0.8)	0.0 (0)
MID	0.0 (0.6)	0.0 (0.0)	1.9 (0.7)	1.5 (0.7)	1.3 (0.6)	0.6 (0.8)	0.0 (0)
SHRUB	0.1 (0.2)	0.1 (0.4)	1.5 (0.8)	1.7 (0.7)	1.6 (0.7)	0.6 (0.8)	0.0 (0)
GC	72.9 (34.0)	80.7 (27.5)	46.4 (37.5)	73.5 (30.1)	1.3 (5.4)	61.3 (40.4)	94.8 (14.6)
GCHT	20.2 (36.2)	14.4 (7.9)	6.9 (8.4)	14.9 (8.8)	1.4 (6.7)	20.4 (12.2)	22.2 (7.4)
NDBH	4.4 (6.1)	4.4 (6.1)	6.8 (3.3)	6.8 (3.3)	8.5 (4.2)	7.0 (8.0)	1.4 (1.9)
DBH	4.5 (6.0)	4.5 (6.0)	11.2 (4.1)	11.2 (4.1)	9.4 (3.7)	7.6 (6.5)	13.7 (15.0)
NSPP	1.3 (1.5)	1.3 (1.5)	3.1 (1.7)	3.1 (1.7)	3.9 (1.4)	2.3 (1.9)	0.7 (0.8)
PCCAN	8.7 (18.9)	12.2 (25.2)	15.4 (11.8)	53.1 (26.7)	23.7 (20.7)	11.4 (17.4)	0.0 (0)

vegetation in the above ground strata. Grid DFI had a more luxurient ground cover than grid DFII. The Flooded grid was a mosaic of forest patches and open field. Canopy height was the most variable of the five grids as was density of vegetation in the three forest strata. Ground cover was dense but highly variable. The Savanna grid was dominated by dense ground cover and little arborescent vegetation except for palms. Habitat profile varied seasonally. The two grids sampled (Dry and DFI) during the wet and dry seasons showed greater canopy and ground coverage in the wet season.

TABLE 13.—*Habitat profiles of small mammals on the Dry grid during wet and dry seasons. Species abbreviations given in Table 4, N=sample size, a prime sign identifies habitat variables used in discriminant analyses, *P<0.05), **P<0.01, ***P<0.001, ns=not significant. Probabilities next to individual habitat variables are the probabilities resulting from one-way ANOVA's between that small mammal profile variable and available habitat. Wilk's Lambda, its associated χ^2, and probability values from the discriminant analysis of species profiles and available habitat are given at the bottom.*

Variable	Wet season			Dry season		
	D. m.	M. r.	Z. b.	D. m.	M. r.	Z. b.
N	26	15	28	13	18	2
CANHT'	5.3 (3.2)***	4.6 (2.4)***	2.3 (3.1)	5.6 (3.4)	5.5 (1.8)***	0.0 (0)
CAN'	1.5 (1.4)***	1.5 (0.9)***	0.6 (0.9)	1.7 (1.2)	1.3 (0.8)***	0.0 (0)
MID'	0.0 (0)	0.0 (0)	0.0 (0)	0.0 (0)	0.0 (0)	0.0 (0)
SHRUB'	0.3 (0.6)*	0.3 (0.5)	0.1 (0.3)	0.0 (0)	0.0 (0)	0.0 (0)
GC'	44.7 (45.8)***	54.0 (39.6)**	77.0 (31.0)	56.5 (34.6)	41.7 (35.2)***	80.0 (28.3)
GCHT'	11.9 (6.0)	11.7 (6.9)	16.9 (10.1)	15.5 (10.0)	8.3 (7.3)	25.5 (6.4)
NDBH'	8.5 (6.7)**	10.1 (5.6)***	7.8 (8.1)*	4.7 (3.5)	12.4 (6.3)***	0.0 (0)
DBH'	9.6 (7.3)***	8.3 (2.6)*	5.2 (4.5)*	8.5 (2.8)*	8.4 (2.4)**	0.0 (0)
NSPP	2.8 (1.3)***	3.2 (1.1)***	1.8 (1.6)	2.5 (0.9)*	1.6 (1.1)***	0.0 (0)
PCCAN	48.5 (41.2)***	46.7 (27.9)***	21.1 (28.9)	50.4 (36.9)	24.4 (22.0)**	0.0 (0)
Lambda	0.71	0.83	0.91	0.67	0.72	0.98
χ^2	41.7	21.2	10.9	43.4	37.5	1.8
P	***	**	ns	***	***	ns

TABLE 14.—*Habitat profiles of small mammals on grid DFI during wet and dry seasons. Abbreviations given in Table 13.*

Variable	Wet season			Dry season		
	M. r.	*Z. b.*	*H. a.*	*M. r.*	*Z. b.*	*H. a.*
N	20	22	14	74	7	26
CANHT′	10.5 (4.1)	8.2 (5.9)	9.6 (4.4)	10.2 (3.5)	8.6 (5.3)	9.8 (4.4)
CAN′	1.6 (0.6)	1.0 (0.6)**	1.2 (0.7)	1.1 (0.5)	0.6 (0.5)	1.0 (0.5)
MID′	1.6 (0.7)	1.5 (0.8)	1.4 (0.9)	1.9 (0.7)	2.3 (0.7)	1.9 (0.6)
SHRUB′	1.7 (0.5)	2.3 (0.8)***	1.9 (0.7)	1.6 (0.9)	2.6 (1.4)**	1.3 (0.5)
GC′	62.0 (32.7)	68.9 (24.9)	58.6 (28.9)*	37.0 (37.1)	40.0 (30.5)	34.4 (37.9)
GCHT′	16.0 (8.7)	19.1 (9.0)	17.9 (10.2)	6.6 (8.6)	19.4 (17.1)***	3.7 (5.7)
NDBH′	7.5 (3.9)	6.1 (3.7)	8.5 (3.7)	8.1 (4.2)*	6.6 (5.2)	6.5 (3.2)
DBH′	10.5 (3.1)	10.3 (4.3)	11.1 (4.4)	9.9 (3.1)*	11.3 (6.7)	11.3 (4.2)
NSPP	3.4 (1.4)	2.9 (1.4)	3.3 (1.3)	3.1 (1.3)	2.7 (0.8)	3.1 (1.4)
PCCAN	65.5 (27.9)*	33.4 (30.2)**	41.4 (31.3)	11.3 (10.0)	7.1 (6.4)	15.2 (12.3)
Lambda	0.93	0.85	0.91	0.93	0.84	0.91
χ^2	7.8	17.9	10.1	12.3	16.9	11.0
P	ns	*	ns	ns	*	ns

Analysis of small mammal microhabitat was limited to species with more than 25 total captures on a grid (Tables 13 to 15). As would be expected, the scansorial species, *Didelphis* and *Marmosa*, were captured more frequently in trap sites in or near areas of dense arborescent vegetation cover. *Zygodontomys* captures were most frequent in areas with dense ground cover. *Heteromys* was captured most frequently in areas of sparse ground cover.

Microhabitat distributions of the different species of small mammals (Table 13 to 15) provide relative indications of habitat similarities or differences among taxa but do not indicate whether each species is actually selecting its microhabitat type from the available habitat. I assume that capture points for a habitat generalist would be distributed randomly throughout a study grid, and comparison of habitat profiles of a generalist species with the available habitat (Table 12) would yield no significant differences in all habitat aspects. Conversely, a specialist would limit its distribution to one specific microhabitat type, which should differ from available habitat. The univariate statistical test of habitat preference tests the null hypothesis that the mean habitat profile per variable for a given species on a specific grid is equal to the mean of that variable over all sampling points on the same grid. The multivariate test of this null hypothesis would concurrently test the equivalence of all habitat variables of available habitat with all variables from the capture profile of the species under examination. The results of one-way ANOVA comparing each variable of available habitat with each variable of species capture profiles are given in Tables 13 to 15. The results of discriminant analysis, which tested the multivariate equivalence of available habitat and species profile, also are given in Tables 13 to 15.

TABLE 15.—*Habitat profile of small mammals on the Flooded grid. Abbreviations are given in Table 13.*

Variable	M. r.	Z. b.
N	41	29
CANHT'	4.4 (4.3)	3.6 (3.3)
CAN'	0.9 (0.8)	1.1 (0.9)*
MID'	0.8 (0.8)	0.7 (0.7)
SHRUB'	1.0 (0.9)*	0.7 (0.9)
GC'	41.9 (41.5)*	63.9 (34.0)
GCHT'	14.8 (8.7)**	21.4 (9.4)
NDBH'	9.1 (8.3)	6.7 (9.7)
DBH'	6.9 (5.2)	8.3 (6.5)
NSPP	2.7 (2.1)	2.1 (1.6)
PCCAN	13.4 (16.9)*	16.2 (19.5)
Lambda	0.91	0.87
χ^2	13.2	16.3
P	ns	*

Zygodontomy was a habitat specialist on grids DFI and Flooded. On all grids on which this species was captured, the preferred microhabitat included areas of dense herbaceous cover and sparse arborescent vegetation. As expected, those two aspects of forest physiognomy are functionally related. As aboveground vegetation density increases, less light penetrates to the ground layer, thereby reducing herbaceous growth. *Heteromys* was a habitat generalist on grid DFI; the habitat profile of capture points did not differ from that of available habitat. *Didelphis* on the Dry grid limited its distribution to the forested portions of the grid. The distribution of *Marmosa* on the Dry grid was similar to that of *Didelphis* and was significantly restricted to wooded capture sites. *Marmosa* was captured more frequently in forested portions of the Flooded grid; however, when all variables were considered simultaneously, there was no significant difference between *Marmosa* microhabitat and available habitat. *Marmosa* did not show any preference for a specific habitat type on grid DFI.

Marmosa and *Zygodontomys* occurred on all grids sampled except the Savanna grid. Overall habitat preferences for those two species were determined by pooling grids Dry, Flooded, and DFI and comparing available habitat to habitat-characterizing capture sites. Inasmuch as available habitat varied seasonally, I only used dry season captures and dry season habitat data. The results (Table 16) indicate that both species are habitat specialists over the broad spectrum of the llanos environment. *Marmosa* captures were most frequent in areas of dense tree and shrub vegetation. *Zygodontomys* captures were greatest in areas of tall ground cover.

TABLE 16.—*Dry season habitat and capture profiles of* Marmosa *and* Zygodontomys. *Habitat and capture data for the grids Dry, DFI, and Flooded used in the analysis. Abbreviations and symbols are defined in Table 13.*

Variable	Habitat (Pooled grids)	Species	
		M. r.	Z. b.
N	296	133	38
CANHT'	5.0 (5.3)	7.8 (4.5)***	4.4 (4.2)
CAN'	0.7 (0.7)	1.0 (0.7)***	0.9 (0.9)
MID'	0.8 (1.0)	1.3 (1.0)***	0.9 (0.9)
SHRUB'	0.7 (0.9)	1.2 (0.9)***	1.0 (1.2)
GC'	60.4 (38.8)	39.2 (38.0)***	60.4 (34.0)
GCHT'	15.9 (23.5)	9.4 (9.2)***	21.3 (10.8)
NDBH'	6.1 (6.2)	9.0 (6.2)***	6.4 (8.9)
DBH'	7.7 (6.3)	8.8 (4.0)	8.5 (6.7)
NSPP	2.2 (1.7)	3.0 (1.5)***	2.1 (1.6)
PCCAN	11.8 (16.5)	13.7 (14.9)	13.7 (17.8)
Lambda		0.89	0.95
χ^2		49.4	18.6
P		***	*

Within each species and grid, I tested the equivalence of microhabitat between sexes, ages, and season. Those tests were done using two-group discriminant analysis. The groups examined were male *vs.* female, adult *vs.* subadult, and captures occurring in the wet season *vs.* dry season. Wilks λ and associated probability of the resulting discriminant function was interpreted as the statistical test of equivalence of microhabitat between the two groups. There were no differences evident between the sexes of any species. There were seasonal differences in microhabitat in *Heteromys* on grid DFI ($\lambda=0.31$, $\chi^2=39.6$, $P<0.001$, $N=26$ dry, 14 wet). The high loadings on the variables GCHT and CAN of the standardized discriminant scores indicated that most of the variation between groups was explained by those variables. *Heteromys* capture sites were characterized by taller ground cover in the wet season (7.9±10.2 cm.) than in the dry season (3.7±5.7 cm.).

There was a significant seasonal difference in *Zygodontomys* microhabitat on the Flooded grid ($\lambda=0.25$, $\chi^2=31.7$, $P<0.001$, $N=8$ dry, 18 wet). The variable SHRUB explained most of the between group variation. *Zygodontomys* captures in the wet season were limited to trap stations with reduced shrub and small tree cover ($\bar{X}$ SHRUB=0.22±0.55) whereas captures in the wet season were characterized by a substantially denser shrub cover ($\bar{X}$ =1.5±0.93). This might reflect movements by *Zygodontomys* to higher and drier portions of the grid immediately following heavy rainfall and subsequent flooding in June. Trees and shrubs often were situated on ground slightly higher than that of surrounding areas of herbaceous vegetation and remained above standing water for much of the wet season.

Microhabitat differences between adult and subadult individuals for *Zygodontomys* occurred on the Dry grid ($\lambda=0.41$, $\chi^2=20.9$, $P<0.01$, $N=24$ adults, 5 subadults). Most of the variation in separating age groups in *Zygodontomys* was explained by the variable NDBH. Subadult animals were captured most frequently in areas containing numerous shrubs and trees ($\bar{X}$ NDBH=18.6±4.9) as compared to adults ($\bar{X}$ NDBH=5.6±6.8). It is possible that the difference in microhabitat between adult and subadult *Zygodontomys* results either from displacement of subadults into suboptimal habitat patches by behaviorally dominant adults or sampling effect due to the low number of subadults captured.

Analysis of distribution of species captures with regard to tree-trap profile was limited to *Marmosa* on grids Dry, DFI, and Flooded and *Didelphis* on the Dry grid (Table 17). Captures of *Marmosa* on the grids DFI and Dry did not differ from available tree trap habitat. However, there were several differences in *Marmosa* habitat relative to available microhabitat on the Flooded grid. Sixty per cent of all *Marmosa* captures on that grid were in traps placed on lianas (TRAPON) whereas only 29 per cent of all the total tree traps available to *Marmosa* were set on lianas. Ninety-three per cent of all *Marmosa* captures were in traps set in shrubs whereas only 73 per cent of the traps available were set on shrubs. All *Marmosa* captures were in traps

TABLE 17.—*Description of tree-trap locations and the distribution of tree trap captures for* Marmosa *and* Didelphis. *Values presented are the number of traps or captures in each category for discrete variables or means (standard deviation in parentheses) for continuous variables. Asterisks indicate levels of probability (see Table 13) resulting from χ^2 tests (for discrete variables) or Wilcoxon two-sample tests for continuous variables between capture profile per species and the distribution of traps per variable class.*

Variable	Grid variables			Species profile			
	Dry	DFI	Flooded	D. m. (Dry)	M. r. (Dry)	M. r. (DFI)	M. r. (Flooded)
N	34	45	48	36	27	73	37
TRAPON							
branch	33	31	36	29*	27	61	15***
liana		13	11			12	22
trunk	1		1	7			
log		1					
TRAPSUP							
tree	34	29	10	36	27	59	1***
shrub		2	27			2	14
log		1					
liana		13	11			12	22
DEDALV							
alive	32	41	43	28*	27	68	37*
dead	2	4	5	8		5	
BRANCH2							
tree	30	38	41	35	26	61	37*
space	4	1	4	1	1		
ground		3	3			10	
shrub		2				2	
trunk		1					
TOPALM							
no	34	13		36	27	10	
yes		30				55	
?		2	48			8	37
TRAPHT	141 (29)	129 (26)	145 (33)	153 (38)	137 (38)	134 (23)	112 (31)
ANGLE	16 (13)	17 (16)	17 (15)	11 (11)	14 (12)	17 (14)	25 (18)*
CANHT	5.9 (2.6)	7.7 (2.8)	5.6 (1.8)	7.9 (4.0)	5.6 (2.9)	7.3 (2.8)	5.3 (1.9)
PALMD	4.0 (0.0)	1.6 (0.7)	1.3 (0.5)	4.0 (0.0)	4.0 (0.0)	1.7 (0.8)	1.6 (0.9)
FIGD	3.8 (0.7)	2.6 (0.8)	1.9 (0.8)	3.3 (1.3)	4.0 (0.0)	2.7 (0.7)	2.1 (0.8)

set in living vegetation that led into the forest canopy rather than abruptly ending in space or leading to the ground. *Marmosa* capture sites were in traps set lower in the vegetation.

Didelphis capture sites differed from available habitat in a number of respects. Whereas only three per cent of available traps were set on tree trunks, 20 per cent of *Didelphis* captures were on trunks. Twenty-two per cent of all captures were on dead vegetation although only six per cent of all tree traps were set in dead vegetation. The mean height of traps that captured *Didelphis* was significantly higher than the mean height of all traps. Captures were most frequent in areas with a tall canopy.

DISCUSSION

There was considerable variation in recapture rates between and within species over the five grids sampled. Generally, marsupials were recaptured more often than were rodents, which was consistent with Fleming's (1971, 1972) observations. There were considerably more captures of male *Marmosa* than of females on the Dry and Flooded grids; however, sex ratios were equal on the other sites. Other investigators have noted a male bias in *M. robinsoni* populations (Fleming, 1972; O'Connell, 1979). The between-grid variation in *Marmosa* sex ratios observed in this study suggests that a numerical dominance of males is not common to all *M. robinsoni* subpopulations. This variation could be a sampling artifact or a habitat specific phenomenon. It is interesting to note that the two grids showing a male bias were both characterized by patchy distributions of arborescent vegetation (August, 1981). Given the scansorial habit of *Marmosa*, its distribution is contingent on the presence of trees and shrubs. Habitats containing disjunct patches of trees and shrubs would have less arborescent vegetation per unit area and might be considered suboptimal habitat. The data on spacing patterns indicated that males had relatively nonoverlapping home ranges. If those ranges represent defended territories, young or behaviorally subordinate males might be displaced into suboptimal habitat. That would result in male-biased sex ratios in dispersal sinks, as has been observed for numerous microtine rodents (Gaines and McClenaghan, 1980). Skewed sex ratios of adult *Marmosa* likely do not result from a production of a greater number of male neonates. Litters typically had a 1:1 correspondence of sexes (Fleming, 1973). Given the data at hand, the above is only a tentative hypothesis. Detailed studies of *Marmosa* movements and spacing are needed to determine if this species is indeed territorial and if subordinate males disperse into suboptimal habitat.

The one-per cent trap success attained in this study is extremely low for small mammal trapping in any habitat. Representative levels of trap success in other environments are 45 per cent in semiarid shrubsteppe of Chile (Pefaur *et al.*, 1979), 34.7 per cent (Cheeseman and Delaney, 1979) and 5.7 per cent (Delaney and Roberts, 1978) in African savanna, 32 per cent for cold desert habitat (O'Farrell *et al.*, 1975), 13.7 per cent in sand dune habitat (Brown, 1973), and 6.8 per cent in Chihuahuan desert (August *et al.*, 1979). The poor trap success in the present study could be due to a low density of small mammals on all of the grids. Other factors might be involved as well. Some species, notably *Echimys* and *Dasyprocta*, became especially trap shy after being captured (August, 1981). Trap disturbance by ants (*Solenopsis* sp.) and termites was common on all grids. Bait removal by leafcutter ants (*Atta* sp.) on the Dry grid was especially troublesome. Future investigators of small mammal ecology in tropical habitats might consider adding insect repellents to baits to reduce this source of disturbance (Anderson and Ohmart, 1977).

Small mammal densities in all habitats were quite low compared to other studies in the Neotropics (Rood and Test, 1968; Fulk, 1975; Dalby, 1975; Fleming, 1975; Pefaur *et al.*, 1979). Severe wet season flooding in my study area probably limits the diversity and abundance of small mammal populations, especially terrestrial species. Evidence in support of this is the habitat shift shown by *Zygodontomys* on the Flooded grid immediately after the start of the rainy season. Dry season habitat was open field that floods in the wet season, whereas wet season microhabitat was shrubby sites. Shrubs usually grow on soil mounds slightly above the wet season flood line and such habitats would afford *Zygodontomys* drier den sites. Worth *et al.* (1968) also found *Zygodontomys* captures increased with the onset of the rainy season. It is interesting to note that *Zygodontomys* is an excellent swimmer and easily crosses flooded areas (personal observation).

Extensive flooding also would affect scansorial mammals. However, because much of their activity is within the vegetation, they might be affected less than terrestrial species. The scansorial *Marmosa robinsoni* occurred in densities greater than those of the terrestrial *Zygodontomys* on all grids. The Savanna grid floods extensively in the wet season, and it is not surprising that the only species of small mammal present was the arboreal *Oryzomys bicolor*.

Wet season flooding might not be the only factor limiting small mammal populations in the llanos. In cultivated rice fields only 10 kilometers from my study area, rodents were a serious pest. Two species, *Holochilus brasiliensis* and *Zygodontomys brevicauda*, would reach densities of hundreds of individuals per hectare (J. Gomez-Nuñez, personal communication). Similar densities of *Zygodontomys* have been reported in other habitats (Karimi *et al.*, 1976). A major difference between cultivated fields and natural habitat for *Zygodontomys* is the predictability and abundance of food resources. The low, dry season productivity of herbaceous vegetation in natural habitats (Monasterio and Sarmiento, 1976) might impose a low carrying capacity for *Zygodontomys*. Grazing by cattle and horses probably has decreased grass biomass that otherwise would have been available to small mammals on Masaguaral. The densities of *Zygodontomys* reported here are not representative for all years. Eight months previous to the initiation of my study, *Zygodontomys* in the vicinity of the Flooded grid occurred at densities that were an order of magnitude greater than I recorded (O'Connell, personal communication).

The distribution and abundance of *Marmosa* among study grids paralleled the presence of optimal habitat. A minimal habitat requisite of this scansorial species is the presence of arborescent vegetation. With the exception of grid DFII, *Marmosa* density was positively correlated with the development of tree and shrub strata. Grid DFII had unusually low densities of small mammals, which might have been due to extreme predation pressures (August, 1981).

Marmosa, Sciurus, and *Didelphis* occurred throughout the range of habitats examined except the Savanna grid. Although *Sciurus* was not captured on the Flooded grid, it is known to occur there (Eisenberg *et al.*, 1979). The other small mammal species showed more restricted distributions. *Oryzomys* was the only small mammal found in the savanna and was infrequently captured in the other habitats. The two arboreal rodents, *Echimys* and *Rhipidomys*, had mutually exclusive ranges; the former preferring deciduous forest; the latter, bajío habitat of the Flooded grid. *Sigmomys* was captured in open field habitat of the Dry and Flooded grids but was absent from deciduous forest and savanna habitats. *Heteromys* was common in the deciduous forest but was uncommon or absent from the other grids (Eisenberg *et al.*, 1979). The distribution and abundance of prime habitat explains some of the variation in species distributions; however, a number of questions exist. Why did most grids support only one species of arboreal rodent? Is the apparent habitat separation of those taxa a product of present or past competition (Weins, 1977)? These questions will remain unanswered until additional field data have been accumulated.

The seasonal variation in small mammal density shows the effects of seasonal rainfall on small mammal populations. *Zygodontomys* captures increased sharply following the start of the wet season. That was not solely due to recruitment of young because breeding was aseasonal; it more likely was due to the dispersal of animals into drier habitat. Karimi *et al.* (1976) found a sharp increase in the density of *Zygodontomys lasiurus* immediately following the start of the rainy season in northeastern Brazil. Their results seem to indicate that the wet season peak in density corresponded to reproduction and weaning of young rather than dispersal.

Marmosa showed a marked lowering of capture frequency and density in the wet season. It is not apparent whether this is due to high mortality or trap shyness. O'Connell (1979) reported a similar pattern of *Marmosa* density in her studies on Masaguaral. *Didelphis* populations declined throughout the study. Other mammalogists working in the same area noted a marked decline in the frequency of *Didelphis* sightings in 1977 and 1978. The causes of that decline in opossum populations is unknown.

The reproductive data indicate that *Marmosa robinsoni* is a seasonal breeder; courtship and copulation occur at the end of the dry season, and young are weaned throughout the wet season. These observations are consistent with those made by O'Connell (1979), Fleming (1973) and Enders (1935, 1966). *Zygodontomys* showed little seasonal pattern in breeding activity. Fleming (1970), Worth *et al.* (1968), Worth (1967), and Enders (1935) reported the same; however, Karimi *et al.* (1976) found seasonal reproduction in *Z. lasiurus.* Considering the energetic demands of pregnancy and lactation, we would predict that breeding would occur during that period of the year when food resources are most abundant (Morse, 1980). *Marmosa* eats insects and fruit (Enders, 1935; Fleming, 1972), and *Zygodontomys* is a herbivore (Fleming, 1970; Walker, 1975). Although I do not have data on the seasonality of insect density on Masaguaral, others (Janzen and Schoener, 1968; Wolda, 1978) have shown a clear wet season increase in insect abundance in tropical Central America. If such a pattern exists in the llanos, *Marmosa* young would be weaned in a resource rich time of the year. Monasterio and Sarmiento (1976) have conducted extensive studies on the phenological patterns of llanos plants. Among herbs and shrubs, most vegetative growth and seed set occurs in the wet season. Considering food resource abundance alone, *Zygodontomys* should restrict breeding activities to the wet season. However, extensive flooding in the wet season likely puts a premium on dry burrow sites for small terrestrial rodents. Therefore, spatial resources might be at their lowest while food resources are highest. Considering both food and space resources, there seems to be no optimal breeding period for *Zygodontomys*, a condition that might explain partially the aseasonal reproductive pattern observed in this study.

There are no general trends in tropical small mammal breeding patterns. African savanna rodents show a wet season breeding peak (Cheeseman and Delaney, 1979). *Heteromys* occurring in wet montane forest breed in the wet season (Rood and Test, 1968). *Liomys salvini* in dry tropical forest of Panamá breed in the dry season (Fleming, 1971). Breeding patterns in tropical small mammals are affected by a multitude of factors, and there are too little comparative data presently available to discern broad similarities in life history strategies (Fleming, 1975).

Teat number and litter size are quite variable in *Marmosa robinsoni*. Enders (1935), Fleming (1973), and Godfrey (1975) reported maximum litter sizes of 13. Enders (1966) found the maximum litter size to be 14, and Tate (1933) reported examining a specimen with 19 attached neonates. Reported teat numbers are 13 (Enders, 1935; Godfrey, 1975), 15 (Enders, 1966), and 19 (Tate, 1933). All *Marmosa* neonates that I observed were attached to teats and this implies that the llanos population of *M. robinsoni* has a larger litter size and teat number (16) than most other populations. A large initial litter size might represent an adaptation to compensate for high mortality imposed by fluctuating resource levels in the extremely seasonal llanos environment (O'Connell, 1979).

The distance moved between captures for *Marmosa* and *Heteromys* was similar to the movement data reported by Fleming (1970, 1972). There is an interesting pattern in the distance moved between successive captures and the dispersion of resources for *Marmosa* and *Zygodontomys*. The grids DFI, Dry, and Flooded show a gradation in the abundance and patchiness of field and forest habitats (August, 1981). Grid DFI had continuous tree and shrub cover and sparse cover of herbaceous vegetation. The Dry and Flooded grids had extensive patches of field mixed with patches of trees and shrubs (matas). Arborescent vegetation was more developed on the Flooded grid as compared to the Dry grid. One would predict that in homogeneous optimal habitat, an animal would need to cover less distance in search of food and nest sites than would an animal occurring in heterogeneous (patchy) optimal habitat. *Marmosa* moved less on the richly wooded grids DFI and Flooded than on the sparsely wooded Dry grid. *Zygodontomys* movements were lowest on the two grids containing extensive patches of preferred open field habitat (Flooded and Dry) as compared to grid DFI. Movement patterns should be correlated with home range size. Indeed, *Marmosa* home range size was smallest in optimal habitat (grid DFI) and largest on the two grids where preferred habitat was patchily distributed (grids Flooded and Dry). Likewise, *Zygodontomys* home range was large in suboptimal habitat (grid DFI) and smaller in preferred habitat (Dry grid). These observations are consistent with the predictions made above, and much of the variation in movements and home range size is explained by the abundance and dispersion of favored habitat (Weins, 1976).

Hunsaker (1977) used Fleming's (1972) movement data to extrapolate home range sizes for *Marmosa* and *Didelphis*. He assumed circular home ranges, the diameter of which equalled the average distance moved between successive captures. Fleming did not calculate home range size so it is impossible to evaluate Hunsaker's estimates. However, I have independent measures of distance moved and home range size and can assess the accuracy of Hunsaker's method of home range extrapolation. Predicted home ranges were calculated by Π(average distanced moved/2)^2. For these calculations, I pooled all grids to increase the sample size of movement values and observed home range estimates. The predicted home ranges were surprisingly close to the observed values. The predicted (first value) and observed (second value) home range (in hectares) for the species tested were: *Zygodontomys* 0.09, 0.11; *Marmosa* 0.17, 0.17; *Didelphis* 0.29, 0.20; and *Heteromys* 0.08, 0.02. Hunsaker's estimate of home range size of 0.20 ha. for *Marmosa* is quite close to my observed value, however, his estimate of *Didelphis* home range (0.43 ha.) was double my observed home range size. Both our estimates of *Didelphis* home range size are probably far too low. This is due to the inability to record long distance movements on small trapping grids. Evidence for this is the strong correlation between estimates of distance moved between successive captures and the intertrap distance chosen by the investigator (August, 1981).

The analysis of spacing behavior suggests exclusive home ranges among male *Marmosa*. Fleming (1972) found the opposite; male ranges overlapped one another, but female ranges did not. Both my study and Fleming's suffer from the bias inherent in measuring space utilization from live trap data (Waser and Wiley, 1979). Future studies based on direct observation of animals in the field (perhaps with the aid of night vision goggles) or radio transmitters (Madison, 1980) attached to animals will provide more accurate data on small mammal spacing patterns. Charles-Dominique (1983) successfully utilized radio transmitters to follow the daily movements of the arboreal marsupials *Caluromys* and *Philànder*. His results show considerable home range overlap and no evidence for territoriality.

An important initial step in the analysis of animal habitat associations is the determination of whether a species selects one microhabitat from the available habitat space (Rosenzweig et al., 1975; Partridge, 1978). Many studies of habitat use in small mammals fail to make this test (Douglas, 1976; M'Closkey, 1976; Dueser and Shugart, 1978, 1979). In comparing habitat utilization per species against all available habitat types present, I found Didelphis and Marmosa to utilize a specific subset of available habitat on the Dry grid but not on grids DFI or Flooded. Similarly, Zygodontomys was a habitat specialist on grids DFI and Flooded but showed no habitat preference on the Dry grid. Why are individuals on one grid habitat specialists whereas individuals on another are habitat generalists? The answer lies in the abundance and distribution of preferred habitat (Weins, 1976). The basic habitat requirement for Marmosa is arborescent vegetation. The preferred habitat of Zygodontomys is dense herbaceous vegetation. Zygodontomys is a habitat generalist on the grid that is predominantly ideal habitat type (Dry grid). When preferred habitat is distributed patchily (DFI and Flooded grids), Zygodontomys restricts its activity to patches of thick herbaceous growth. The same pattern is evident for Marmosa. Grid DFI was for the most part optimal Marmosa habitat, hence habitat selection was not observed. On the Dry grid, trees and shrubs formed discrete patches within field habitat, and it was in these patches that Marmosa restricted its activity. I would have expected greater habitat specialization by Marmosa on the Flooded grid, but the marginally significant ($P=0.10$) difference between Marmosa habitat profile and available habitat indicates a trend toward specialization.

There was no difference in habitat utilization between sexes in any species. The only difference in habitat preference between adult and subadult animals was for Didelphis on the Dry grid; subadults were captured more frequently at trap stations with a higher canopy and greater density of trees and shrubs. All small mammal species on grid DFI showed seasonal differences in habitat utilization. This is not so much a shift in habitat preference, rather it is a product of the dramatic change in habitat profile on that grid between seasons. Zygodontomys showed an interesting spatial niche shift at the onset of the wet season on the Flooded grid. Once flooding began, Zygodontomys captures were restricted to trap stations near shrubby portions of the grid. Those shrub patches were usually above the flood line and would afford that species drier nest sites.

Competition is an important factor in structuring many small mammal communities (Brown, 1975, 1978; Grant, 1978; Dueser and Hallet, 1980; Munger and Brown, 1981). Coexistence is promoted by reducing interspecific overlap in habitat utilization or food resources (Rosenzweig and Winakur, 1969; Brown and Lieberman, 1973; Smigel and Rosenzweig, 1974; Rosenzweig et al., 1975; Dueser and Shugart, 1978, 1979; M'Closkey, 1978). Is there evidence for competition among species of small mammals in the llanos? Rigorous statistical analyses are available to identify potentially competing species dyads (Hallet and Pimm, 1979); however, given the small sample size of many of the taxa that could potentially compete, the results of such analyses would be suspect. It is clear that considerable niche partitioning occurs spatially. I would not expect intense competition among arboreal (*Echimys*, *Rhipidomys*, and *Oryzomys*) and terrestrial (*Zygodontomys*, *Sigmomys*, and *Heteromys*) taxa. Within species groups that occupy similar aspects of the habitat, microhabitat partitioning is apparent (for example, *Didelphis* versus *Marmosa*; *Zygodontomys* versus *Heteromys*). Complete analysis of competitive relationships between llanos small mammals will require reliable data on food habits. Almost nothing is known of the diets of Neotropical small mammals, much less those in the llanos.

In summary, the Venezuelan llanos supports a small mammal fauna of low diversity and density. The seasonality of rainfall in the llanos is extreme and plays a significant role in many aspects of small mammal life history, especially reproduction and habitat selection. The llanos consists of a mosaic of habitats ranging in complexity from simple savannas to structurally complex deciduous forests. The abundance and distribution of habitat type affects the diversity and relative abundance of small mammals. Much of the variation in movements, home range size, and habitat selection within a species occupying different habitat types is explained by the extent and patchiness of preferred habitat.

Acknowledgments

My fieldwork in Venezuela was sponsored by a Predoctoral Fellowship from the Smithsonian Institution. During the data analysis and writing phase of this study, I was supported by Teaching Fellowships from the Department of Biology, Boston University and the Albert R. and Alma Shadle Fellowship from the American Society of Mammalogists. J. Eisenberg and T. Kunz provided expert advice throughout this project. I extend my deepest thanks to Sr. T. Blohm, the owner of Fundo Pecuario Masaguaral, for allowing my wife and me to live and study on his conservation ranch. Logistic support in Venezuela was generously provided by E. Mondolfi, J. Gomez-Nuñez, R. Rudran, M. Carballo, J. Figuroa, M. O'Connell, and B. Howser. T. Kunz, J. Eisenberg, R. Tamarin, and F. Wasserman read the manuscript and made many helpful suggestions. Lynn August assisted in all aspects of this study; her help and support were above and beyond the call of matrimonial duty. The Boston University Academic Computing Center generously provided computing time for data analysis.

Literature Cited

Anderson, B. W., and R. D. Ohmart. 1977. Rodent bait additive which repels insects. J. Mamm., 58:242.

August, P. V. 1981. Population and community ecology of small mammals in northern Venezuela. Unpublished Ph.D. dissertation, Boston University, 218 pp.

AUGUST, P. V., J. CLARKE, M. H. McGAUGH, AND R. L. PACKARD. 1979. Demographic patterns of small mammals, a possible use in impact assessment. Pp. 333-340, *in* Biological investigations in the Guadalupe Mountains National Park (R. J. Baker and H. H. Genoways, eds.). Nat. Park Serv. Proc. and Trans., 4:333-340.

BARNES, R. D. 1977. The special anatomy of *Marmosa robinsoni*. Pp. 387-413, *in* The biology of marsupials (D. Hunsaker, ed.). Academic Press, New York, xv+537 pp.

BARR, A. J., J. H. GOODNIGHT, AND J. P. SALL. 1979. SAS user's guide. SAS Institute Inc., Cary, North Carolina, 494 pp.

BROWN, J. H. 1973. Species diversity of seed-eating desert rodents in sand dune habitats. Ecology, 54:775-787.

———. 1975. Geographical ecology of desert rodents. Pp. 315-341, *in* Ecology and evolution of communities (M. L. Cody and J. M. Diamond, eds.). Belknap Press, Cambridge, Masachusetts, xii+545 pp.

———. 1978. Effects of mammalian competitors on the ecology and evolution of communities. Pp. 52-57, *in* Populations of small mammals under natural ecosystems (D. P. Snyder, ed.). Pymatuning Symposia in Ecology, Pymatuning Laboratory of Ecology, Univ. of Pittsburgh, 5:xiii+1-237.

BROWN, J. H., AND G. A. LIEBERMAN. 1973. Resource utilization and coexistence of seed-eating desert rodents in sand dune habitats. Ecology, 54:788-797.

CHARLES-DOMINIQUE, P. 1983. Ecology and social adaptations in didelphid marsupials: comparison with eutherians of similar ecology. Pp. 395-422, *in* Advances in the study of mammalian behavior (J. F. Eisenberg and D. G. Kleiman, eds.). Spec. Publ. Amer. Soc. Mamm., xvi+ 753 pp.

CHEESEMAN, C. L., AND M. J. DELANEY. 1979. The population dynamics of small rodents in a tropical African grassland. J. Zool., 188:451-475.

COOLEY, W. W., AND P. R. LOHNES. 1971. Multivariate data analysis. John Wiley and Sons, New York, xii+364 pp.

DALBY, P. L. 1975. Biology of Pampa rodents. Publ. Mus. Michigan St. Univ., Biol. Ser., 5:149-272.

DELANEY, M. J., AND C. J. ROBERTS. 1978. Seasonal population changes in rodents in the Kenya Rift Valley. Bull. Carnegie Mus., 6:97-108.

DOUGLASS, R. J. 1976. Spatial interactions and microhabitat selection of two locally sympatric voles, *Microtus montanus* and *Microtus pennsylvanicus*. Ecology, 57:346-352.

DUESER, R. D., AND J. H. HALLET. 1980. Competition and habitat selection in a forest-floor small mammal fauna. Oikos, 35:293-297.

DUESER, R. D., AND H. H. SHUGART. 1978. Microhabitats in a forest-floor small mammal fauna. Ecology, 59:89-98.

———. 1979. Niche pattern in a forest-floor small mammal fauna. Ecology, 60:108-118.

EISENBERG, J. F., M. A. O'CONNELL, AND P. V. AUGUST. 1979. Density, productivity, and distribution of mammals in two Venezuelan habitats. Pp. 187-207, *in* Vertebrate ecology in the northern Neotropics (J. F. Eisenberg, ed.). Smithsonian Institution Press, Washington, D.C., 271 pp.

ENDERS, R. K. 1935. Mammalian life histories from Barro Colorado Island, Panama. Bull. Mus. Comp. Zool., 78:383-502.

———. 1966. Attachment, nursing, and survival of young in some didelphids. Symp. Zool. Soc. London., 15:195-203.

FLEMING, T. H. 1970. Notes on the rodent faunas of two Panamanian forests. J. Mamm., 51:473-490.

———. 1971. Population ecology of three species of Neotropical rodents. Misc. Publ. Mus. Zool. Univ. Michigan, 143:1-77.

———. 1972. Aspects of the population dynamics of three species of opossums in the Panama Canal Zone. J. Mamm., 53:619-623.

————. 1973. The reproductive cycles of three species of opossum and other mammals in the Panama Canal Zone. J. Mamm., 54:439-455.

————. 1975. The role of small mammals in tropical ecosystems. Pp. 269-298, *in* Small mammals: their productivity and population dynamics (F. B. Golley, K. Petrusewicz, and L. Ryskowski, eds.). I.B.P. 5, Cambridge University Press, Cambridge, xxv+451 pp.

FULK, G. W. 1975. Population ecology of rodents in the semiarid shrublands of Chile. Occas. Papers Mus., Texas Tech Univ., 33:1-40.

GAINES, M. S., AND L. R. MCCLENAGHAN. 1980. Dispersal in small mammals. Annual Rev. Ecol. Syst., 11:163-196.

GODFREY, G. K. 1975. A study of oestrus and fecundity in a laboratory colony of mouse opossums (*Marmosa robinsoni*). J. Zool., 175:541-555.

GRANT, P. R. 1978. Competition between species of small mammals. Pp. 38-51, *in* Populations of small mammals under natural ecosystems (D. P. Snyder, ed.). Pymatuning Laboratory of Ecology, Univ. Pittsburgh, 5:xiii+1-237.

GREEN, R. H. 1971. A multivariate statistical approach to the Hutchinsonian niche: bivalve molluscs of central Canada. Ecology, 52:543-556.

————. 1980. Multivariate approaches in ecology: the assessment of ecological similarity. Annual Rev. Ecol. Syst., 11:1-14.

HALLETT, J. G., AND S. L. PIMM. 1979. Direct estimation of competition. Amer. Nat., 113:593-600.

HUNSAKER, D. 1977. The biology of marsupials. Academic Press, New York, xvi+537 pp.

JANZEN, D. H., AND T. W. SCHOENER. 1968. Differences in insect abundance and diversity between wetter and drier sites during a tropical dry season. Ecology, 49:96-110.

KARIMI, Y., C. RODRIGUES DE ALMEIDA, AND F. PETTER. 1976. Note sur les rongeurs du Nord-est du Bresil. Mammalia, 40:257-266.

LACHENBRUCH, P. A. 1975. Discriminant analysis. Hafner, New York, ix+128 pp.

MADISON, D. W. 1980. Space use and social structure in meadow voles, *Microtus pennsylvanicus*. Behav. Ecol. Sociobiol., 7:65-71.

MARES, M. A. 1980. Convergent evolution among desert rodents: a global perspective. Bull. Carnegie Mus. Nat. Hist., 16:1-51.

MARES, M. A., K. E. STREILEIN, AND M. R. WILLIG. 1981. Experimental assessment of several population estimation techniques on an introduced population of eastern chipmunks. J. Mamm., 62:315-328.

M'CLOSKEY, R. T. 1976. Community structure in sympatric rodents. Ecology, 57:728-739.

————. 1978. Niche separation and assembly in four species of Sonoran desert rodents. Amer. Nat., 112:683-694.

MCNAB, B. K. 1963. Bioenergetics and the determination of home range size. Amer. Nat., 97:133-140.

MESERVE, P. L., AND W. E. GLANZ. 1978. Geographical ecology of small mammals in the northern Chilean arid zone. J. Biogeogr., 5:135-148.

MONASTERIO, M., AND G. SARMIENTO. 1976. Phenological strategies of plant species in the tropical savanna and the semi-deciduous forest of the Venezuelan llanos. J. Biogeogr., 3:325-356.

MORSE, D. H. 1980. Behavioral mechanisms in ecology. Harvard Univ. Press, Cambridge, Massachusetts, viii+383 pp.

MUNGER, J. C., AND J. H. BROWN. 1981. Competition in desert rodents: an experiment with semipermeable exclosures. Science, 211:510-512.

NEFF, N. A., AND L. MARCUS. 1980. A survey of multivariate methods for systematics. New York (privately published) x+243 pp.

NEFF, N. A., AND G. R. SMITH. 1979. Multivariate analysis of hybrid fishes. Syst. Zool., 28:176-196.

NIE, N. H., C. H. HYLL, K. STEINBRENNER, AND D. H. BENT. 1975. SPSS, statistical package for the social sciences. McGraw Hill, New York, xxiv+675 pp.

O'CONNELL, M. A. 1979. Ecology of didelphid marsupials from northern Venezuela. Pp. 73-87, in Vertebrate ecology in the northern Neotropics (J. F. Eisenberg, ed). Smithsonian Institution Press, Washington, D.C., 271 pp.

O'FARRELL, T. P., R. J. OLSON, R. O. GILBERT, and J. D. HEDLUND. 1975. A population of great basin pocket mice, Perognathus parvus, in the shrub-steppe of south-central Washington. Ecol. Monog., 45:1-28.

PARTRIDGE, L. 1978. Habitat selection. Pp. 351-376, in Behavioural ecology (J. R. Krebs and N. B. Davies, eds.). Sinauer Assoc., Sunderland, Massachusetts, x+494 pp.

PEARSON, O. P. 1975. An outbreak of mice in the coastal desert of Peru. Mammalia, 39:375-386.

PEFAUR, J. E., J. L. YANEZ, AND F. M. JAKSIC. 1979. Biological and environmental aspects of a mouse outbreak in the semi-arid region of Chile. Mammalia, 43:313-322.

PIZZIMENTI, J. J., AND R. D. DE SALLE. 1980. Dietary and morphometric variation in some Peruvian rodent communities: the effect of feeding strategy on evolution. Biol. J. Linn. Soc., 13:263-285.

RAMIA, M. 1967. Tipos de sabanas en los llanos de Venezuela. Bol. Soc. Venezolana Cienc. Nat., 27:264-288.

ROOD, J. P., AND F. H. TEST. 1968. Ecology of the spiny rat, Heteromys anomalus at Rancho Grande, Venezuela. Amer. Midland Nat., 78:89-102.

ROSENZWEIG, M. L., B. SMIGEL, AND A. KRAFT. 1975. Patterns of food, space, and diversity. Pp. 241-268, in Rodents in desert environments (I. Prakash and P. K. Glosh, eds.). W. Junk, The Hague, xvi+624 pp.

ROSENZWEIG, M. L., AND J. WINAKUR. 1969. Population ecology of desert rodent communities: habitats and environmental complexity. Ecology, 50:558-572.

SARMIENTO, G., AND M. MONASTERIO. 1969. Studies on the savanna vegetation of the Venezuelan llanos. I. The use of association analysis. J. Ecol., 57:579-598.

———. 1971. Ecologia de las sabanas de America tropical. I. Analisis macroecological de los llanos de Calabozo, Venezuela. Cuad. Geogr., 4:1-127.

———. 1975. A critical consideration of environmental conditions associated with the occurrence of savanna ecosystems in tropical America. Pp. 223-250, in Tropical ecological systems (F. B. Golley and E. Medina, eds.). Springer-Verlag, New York, xiii+398 pp.

SMIGEL, B., AND M. L. ROSENZWEIG. 1974. Seed selection in Dipodomys merriami and Perognathus penicillatus. Ecology, 55:329-339.

SOKAL, R. R., AND F. J. ROHLF. 1969. Biometry. W. H. Freeman and Co., San Francisco, California, xxi+776 pp.

TATE, G. H. H. 1933. A systematic revision of the marsupial genus Marmosa. Bull. Amer. Mus. Nat. Hist., 66:1-250.

TROTH, R. G. 1979. Vegetational types on a ranch in the central llanos of Venezuela. Pp. 17-30, in Vertebrate ecology in the northern Neotropics (J. F. Eisenberg, ed.). Smithsonian Institution Press, Washington, D.C., 271 pp.

WALKER, E. P., AND OTHERS. 1975. Mammals of the world, 3rd ed. John Hopkins Press, Baltimore, 2 volumes.

WASER, P. M., AND R. H. WILEY. 1979. Mechanisms and evolution of spacing in animals. Pp. 159-223, in Handbook of neurobiology (P. Marler and J. G. Vandenbergh, eds.). Plenum Press, New York.

WEINS, J. A. 1976. Population responses to patchy environments. Annual Rev. Ecol. Syst., 7:81-120.

———. 1977. On competition and variable environments. Amer. Sci., 65:590-597.

WOLDA, H. 1978. Seasonal fluctuations in rainfall, food, and abundance of tropical insects. J. Animal Ecol., 47:369-381.

WORTH, C. B. 1967. Reproduction, development, and behavior of captive *Oryzomys laticeps* and *Zygodontomys brevicauda* in Trinidad. Lab. Animal Care, 17:355-361.

WORTH, C. B., W. G. DOWNS, T. H. G. AITKEN, AND E. S. TIKASINGH. 1968. Arbovirus studies in Bush Bush Forest, Trinidad, W. I., September 1959-December 1964. IV. Vertebrate populations. Amer. J. Tropic. Med. and Hyg., 17:269-275.

ECOLOGY OF THE SPOTTED GROUND SQUIRREL, SPERMOPHILUS SPILOSOMA (MERRIAM), ON PADRE ISLAND, TEXAS

JAMES C. SEGERS AND BRIAN R. CHAPMAN

Although there have been many ecological and behavioral studies of ground squirrels of the genus *Spermophilus*, there is little ecological information available on the spotted ground squirrel, *S. spilosoma*. The secretive nature of this species, plus its preference for arid areas could account for the scarcity of ecological studies. Sumrell's (1949) report on the life history of the species in New Mexico and the general accounts of Hall (1955), Davis (1974), and Schmidly (1977) are among the few ecological descriptions of the species.

Spotted ground squirrels occur from southern Dakota south to Durango, México, in an area generally west of the 100° meridian and east of the 110° meridian (Hall, 1981). They are common on a variety of substrates ranging from dense clays or gravel to loose sands but are usually associated with sparsely vegetated grasslands or brushlands (Davis 1974). Spotted ground squirrels on Padre Island live in the sand dunes and on the margins of sand-washover zones.

MATERIALS AND METHODS

The primary area used for this study was located on Padre Island, Kleberg County, Texas, 3.2 km. north of the entrance to Padre Island National Seashore. Padre Island is one of a series of Pleistocene barrier islands along the Texas Coast. Hill and Hunter (1976) described landform, climate, and geological history of this area. Judd *et al.* (1977) and Lonard and Judd (1980) described the vegetation of south Padre Island in relation to topography. Four vegetational zones were recognized within the study area. the active dunes (Zone 1) were characterized by moving sand, sparse vegetation, and little relief (2 to 4 meters). Dominant vegetation consisted of Beach Tea (*Croton punctatus* Jacq.), Beach Evening-Primrose (*Oenothera drummondii* Hook), Marshhay Cordgrass (*Spartina patens* (Ait.) Muhl.), and Ground Cherry (*Physalis viscosa* L.). The deflation flat (Zone 2) supported a dense ground cover of tall grasses, the dominant species of which were Seacoast Bluestem (*Schizachyrium scoparium* (Michx.) Nash), Bushy Bluestem (*Andropogon glomeratus* (Watt.) B.S.P.), and Gulf Dune Paspalum (*Paspalum monstachyum* Vasey). Low areas were characterized by *Eleocharis* sp., Marsh Penneywort (*Hydrocotyle bonariensis* Lam.), and Marshhay Cordgrass (*Spartina patens*). The herbaceous area (Zone 3), adjacent to the active dunes had little relief (0.5 to 1 meters) and was sparsely vegetated. Dominant species were Gulf Dune Paspalum (*Paspalum monstachyum*), Marshhay Cordgrass (*Spartina patens*), and two legumes, Wild Indigo (*Baptisia leucophaea* Nutt), and Patridge Pea (*Cassia fasciculata* Michx.). The saturated depression (Zone 4) was characterized by semipermanent standing water; the water table was within a few centimeters of the surface. Dominant vegetation consisted of Yellow Nutgrass (*Cyperus esculentus* L.), and Juncus (*Juncus scirpoides* Lam.). Formation and structure in those areas were described by Hunter and Dickinson (1970).

A second area having the same floristic associations was used to determine food habits and the extent of activity during the months of dormancy (December 1976 through February 1977). That area was located at Nueces County Park, Nueces County, Texas, 12.8 km. north of the primary site.

Data were collected from 1 May 1976 to 1 May 1977. One hundred live traps (similar in size and construction to the Havahart No. 2), were placed at 19-meter intervals in a 10-trap by 10-trap, 4.4 hectare grid, which was positioned 400 meters behind the foredune ridge. A single trap was placed at each station and baited with hen scratch.

Trapping was conducted in two periods daily: the morning period began at sunrise and lasted 4 to 5 hours; the second period began 4 to 5 hours prior to sunset and lasted until sunset. Trapping was conducted weekly on 2 consecutive days. Trap periods were designed to avoid midday heat and to correspond to the crepuscular activity patterns of ground squirrel (Davis, 1974).

Captured ground squirrels were marked by toe clipping (Baumgartner, 1940) and fur dyeing in various patterns; however, the latter technique proved to be unreliable and was stopped after two months. Ground squirrels were examined for reproductive condition at each capture. This position of testes in males was noted and recorded as inguinal or scrotal. Females were examined and their condition was recorded as pregnant, lactating, or no sexual activity. Two age groups were recognized; young of the year and adults.

Trap ranges of all spotted ground squirrels captured during the study were calculated by the inclusive boundary strip method (Stickel, 1954). Those data were separated by age and sex and compared by means of Student's *t*-test.

Vegetation was analyzed from 50-meter line-intercept transects (Benton and Werner, 1972). A total of six transects were run, one each in the deflation flat and saturated depression and two each in the active dunes and herbaceous area. Captured ground squirrels were followed upon release and the location of the burrow they entered recorded on a site map. At the conclusion of all trapping, all burrows (*N*=28) were excavated and diagrammed.

Observation periods on the primary and secondary areas were conducted throughout the study to determine food utilized and the extent of activity during the winter months. Observations on the primary area coincided with trap periods and extended 4 to 5 hours. These periods were conducted twice per month except during the dormant period. Between November 1976 and February 1977 one-hour observations were conducted on both study areas twice weekly at various times of the day.

Results and Discussion

A total of 33 spotted ground squirrels were captured 71 times during 208 trap periods. An additional 105 hours of direct observation were conducted.

Mean population size of ground squirrels captured monthly was 6.5 animals, excluding December 1976 and January 1977. We feel that trap data taken between November 1976 and February 1977 reflected reduced activity due to weather-related dormancy rather than actual population size. The largest monthly population (14), recorded in August 1976, contained 8 adult and 6 young of the year. The emergence of young animals in late summer created the largest yearly populations. Accordingly, the highest densities occurred in August and September, 3.2 and 2.3 animals per hectare. Mean density, excluding the questionable months, was 1.5 per hectare.

Low density of animals presented a problem in estimating accurately the population size. McCarley (1958) described a nonstatistical method of estimating population size, which he defined as the total number of animals caught in a given month plus any animal caught prior to that month or again afterwards. In our study, population was estimated by McCarley's method (Fig. 1).

Davis (1974) reported that breeding occurs between February and July in Texas. Breeding period was determined indirectly by the emergence of young animals. Several investigators (Bridgewater, 1966; Morton and Gallup, 1975) have reported gestation to be 27-28 days in species of *Spermophilus*. Development within the burrow prior to emergence in the Mexican ground squirrel was 35 days (Matocha, 1969). Assuming young animals were caught within 5 days of emergence (the greatest time between trap periods) the breeding period was calculated to be from March to October. The earliest scrotal male was captured 3 March 1977; the latest, 31 August. The earliest pregnant female was captured 21 April and the last lactating female 1 October. No evidence of two litters per year was observed in marked females, and no young of the year were observed in breeding condition.

Males accounted for 51.6 per cent of the population during the period of study. Adult males moved the greatest maximum distances, with a range of 2210.0 square meters; adult females, 1925.3; juvenile males and females, 1203.3 and 1444.0, respectively. Difference in trap range size between adult males and females was significant ($P < 0.05$, df = 8) and might reflect a tendency of males to travel more widely (Chapman and Packard, 1974). Small ranges of juveniles might reflect a tendency for a dam's offspring to stay within her home range.

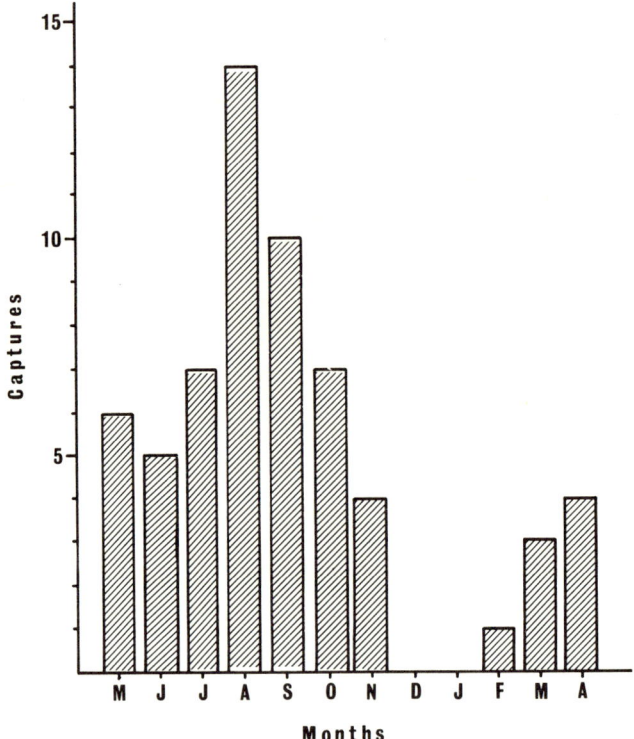

Fig. 1.—Captures per month, a population estimate designed by McCarley (1958).

Vegetation was analyzed in June 1976; cover was calculated for each zone and then compared against trap catches to determine requirements (Table 1). Squirrels most often were captured in the active dunes. In the herbaceous area, squirrels followed established runways to feeding areas or open places. Little activity was recorded by observation and trapping in the two remaining zones, and no squirrels captured were in the deflation flat.

A list of food items utilized by squirrels was compiled during periods of observation (Table 2). Seeds comprised the largest percentage of food except during spring when green plant parts and insects were eaten. Grass seeds were the major food item found stored in burrrows. Occasionally, animals were eaten; squirrels often were observed chasing lizards and insects, and a partially consumed earless lizard (*Holbrookia* sp.) was found in one burrow.

A total of 28 burrows was excavated. Desha (1966) described two burrow types used by thirteen-lined ground squirrels (*S. tridecemlineatus*): refuge burrows and home burrows. Here, refuge burrows were the most frequent type (26 of 28), and they often had multiple entrances (three to six) and single chambers (11 of 26). Entrances to refuge burrows usually were not hidden. Squirrel burrows generally were shallow (10 to 80 centimeters below

TABLE 1.—*Comparison of the number of captures and cover in each vegetative zone.*

Vegetation zone	Cover value (%)	Captures (%)	No. of traps in zone	Capture index*
Active dunes	26.5	82	35	1.70
Herbaceous	93.1	10	34	0.21
Saturated depression	25.7	8	16	0.62
Deflation flat	99.8	0	15	0.00

Capture index is the total captures in each vegetative zone divided by the total number of traps in that zone.

surface) and relatively short (mean, 1.6 m.; range, 0.5 to 6.0). Mean diameter of all tunnels was 7 centimeters (5 to 9)

Home burrows contained food caches and nest materials and were much less numerous (two of 28). Entrances to home burrows (2 to 3) were well hidden by vegetation and tended to be located high in the dunes. Nest chambers contained little vegetation except some grass and leaf litter, but were enclosed with roots of overlying vegetation. Typical home and refuge burrows are shown in Fig. 2.

Habitat-use patterns of S. *spilosoma* differ from those reported for other ground squirrel species (Edwards, 1946; Desha, 1966; McCarley, 1966) in several aspects. Most *Spermophilus* forage in relatively dense cover. Forays for food are usually brief, and the animals return after each foray to the home burrow. The spotted ground squirrel, however, leaves the immediate surrounding of the home burrows to forage extensively in open or barren areas. Such wide ranging activities in barren substrates could reduce the chances of ambush and capture by diurnal snakes (Vaughn and Schwartz, 1980). The construction and use of refuge burrows could permit rapid escape from aerial predators while freeing the squirrels to forage more widely on the perimeter of their territories and to deter encroachment by neighbors.

Spotted ground squirrels appear to be opportunistic feeders, taking food items as they are seasonally available. Their preference for poorly vegetated

TABLE 2.—*Observed plant species utilized for food during the year.*

Species	Food item	Time of year
Poaceae		
Paspalum monostachyum	seedheads	Late summer-autumn
Schizachyrium scoparium	seedheads	Autumn
Spartina patens	stems	Spring-Summer
Uniola paniculata	seeds	Late summer
Fabaceae		
Cassia fasciculata	seeds	Autumn
Onagraceae		
Oenothera drummondi	stems	Spring-summer-autumn
Solanaceae		
Physalis viscosa	fruit	Spring-summer

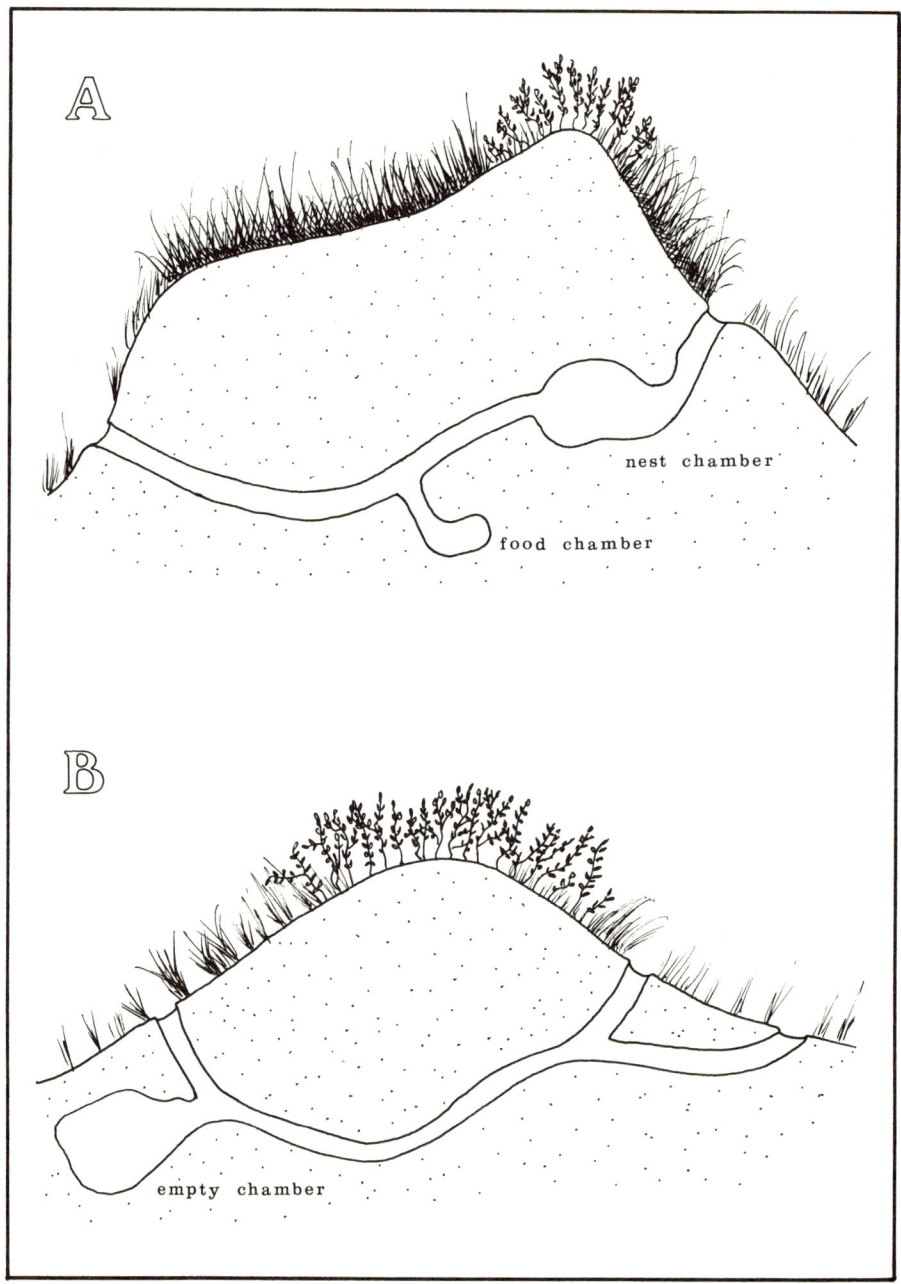

FIG. 2.—Diagram of a typical home burrow (A) and refuge burrow (B). Drawings are not to scale.

areas and their habit of caching food (presumably for the winter) necessitate a large home range. Thus, the low population density observed on the study area might be typical for the entire island.

Acknowledgments

This research was supported in part by a Grant-in-Aid from the Society of Sigma Xi and by a scholarship from the Science Association, Corpus Christi State University. Our special thanks to Drs. Bart Cook, J. W. Tunnell, Jr., John Benedict and two anonymous reviewers for their criticism of this manuscript.

Literature Cited

Baumgartner, L. L. 1940. Trapping, handling and marking fox squirrels. J. Wildlife Manage., 4:444-450.

Benton, A. H., and W. F. Werner, Jr. 1972. Manual of field biology and ecology. Burgess Publishing Co. Minneapolis, Minn., 400 pp.

Bridgwater, D. D. 1966. Laboratory breeding, early growth, development and behavior of *Citellus tridecemlineatus* (Rodentia). Southwestern Nat., 11:325-337.

Chapman, B. R., and R. L. Packard. 1974. An ecological study of Merriam's pocket mouse in southeastern Texas. Southwestern Nat., 19:281-191.

Davis, W. B. 1974. The mammals of Texas. Bull. Texas Parks and Wildlife Dept., 41 (revised): 1-294.

Desha, P. G. 1966. Observations of burrow utilization of the thirteen-lined ground squirrel. Southwestern Nat., 11:408-410.

Edwards, R. L. 1946. Some notes on the Mexican ground squirrel in Texas. J. Mamm., 27:105-115.

Hall, E. R. 1955. The mammals of Kansas. Univ. Kansas Mus. Nat. Hist., Misc. Publ. 7:1-303.

———. 1981. The mammals of North America. 2nd ed. John Wiley and Sons, New York, 1:xviii+1-600+90.

Hill, G. W., and R .E. Hunter. 1976. Interaction of biological and geological processes in the beach and nearshore environments, northern Padre Island, Texas. *in* A. Davis and R. A. Ephington (eds.), Beach and Near-shore Sedimentation, Soc., Econ. Paleontol. Minearol. Spec. Publ., 24:169-187.

Hunter, R. E., and K. A. Dickinson. 1970. Map showing land forms and sedimentary deposits of the Padre Island portion of the South Bird Island 7.5 minute quadrangle, Texas. U.S. Geological Survey.

Judd, F. W., R. I. Lonard, and S. L. Sides. 1977. The vegetation of South Padre Island, Texas in relation to topography. Southwestern Nat., 22:31-48.

Lonard, R. I., and F. W. Judd. 1980. Phytogeography of South Padre Island, Texas. Southwestern Nat., 25:313-322.

Matocha, K. 1969. A study of certain aspects of the reproduction, growth and development of the Mexican ground squirrel (*Citellus mexicanus*) in southern Texas. Unpublished M. S. Thesis. Texas A&I Univ., Kingsville, Texas, 52 pp.

McCarley, H. 1958. Ecology, behavior and population dynamics of *Peromyscus nuttalli* in eastern Texas. Texas J. Sci., 10:147-171.

———. 1966. Annual cycle, population dynamics and adaptive behavior of *Citellus tridecemlineatus*. J. Mamm., 47:294-316.

Morton, M. L., and J. S. Gallup. 1975. Reproductive cycle of the Belding ground squirrel (*Spermophilus beldingi beldingi*): Seasonal and age differences. Great Basin Nat., 25:427-433.

SCHMIDLY, D. J. 1977. The mammals of Trans-Pecos Texas. Texas A&M Univ. Press, College Station, 225 pp.

SMITH, J. R. 1973. Hibernation in the spotted ground squirrel, *Spermophilus spilosoma annectens.* J. Mamm., 34:316-324.

STICKEL, L. F. 1954. A comparison of certain methods of measuring ranges of small mammals. J. Mamm., 35:1-15.

SUMRELL, F. 1949. A life history study of the ground squirrel *Citellus spilosoma major* (Merriam). Unpublished M.S. Thesis., Univ. of New Mexico, Albuquerque, 105 pp.

VAUGHN, T. A., AND S. T. SCHWARTZ. 1980. Behavioral ecology of an insular woodrat. J. Mamm., 61:205-218.

RATES OF INCREASE IN LONG-LIVED ANIMALS WITH A SINGLE YOUNG PER BIRTH EVENT

WALT CONLEY

The potential for increase has been a primary subject of interest to population biolgists since the original essay by Malthus (1798). Even though population parameters that reflect rate of increase are intuitively appealing and necessary for various applications, adequate estimates are difficult to obtain. There are various reasons for such difficulties, which have led to a diverse literature that attempts to provide solutions for both analytical and estimation problems. In this paper, I deal primarily with the analytical and conceptual difficulties associated with estimated rates of increase in animal populations. Much of the discussion to follow is general in that it applies to any population with similar life-history patterns. My purpose is to establish a series of theoretical upper boundaries for rates of increase in such populations. At the same time, I have provided a series of graphs that represent various combinations of the primary population parameters of mortality and natality, in an effort to clarify the interactions that result in various observed rates of increase.

A theoretical approach of this form establishes certain biological and mathematical "rules of the game." Additionally, given agreement on basics, a foundation is thus provided for the interpretation, and judgment of the worth, of available information pertaining to real populations.

The available literature on mathematical demography ranges from highly sophisticated and abstract to superficial and perhaps trivial. Within this body of literature, presentations exist that are intermediate in mathematical complexity, and that are intended to facilitate understanding of the biological processes involved (for example, Caughley, 1966, 1967a, 1967b, 1977; Cole, 1954; Conley, 1978; Eberhardt, 1969; Mertz, 1970). Throughout, however, there are major problems involved with the appropriate estimation of demographic parameters from wild populations (see Pospahala et al., 1974; Anderson, 1975).

Although problems of estimation are real, and associated implicit assumptions are highly restrictive, these two facts do not excuse the continued misuse of demographic techniques, nor do they excuse statements regarding "realized" rates of increase that are impossible even in a theoretical context, much less in the real populations presumably being described.

It is not my purpose here to contribute to a discussion of the various interpretations available concerning "rates-of-increase" in populations. As a result, I have avoided such terms as "intrinsic rate of increase," "r-max," "Malthusian rate of increase," and the like. For an introduction to the discussion that follows, see Caughley (1977), Conley (1978), Conley and Nichols (1978), and Slade and Balph (1974).

METHODS

Specification of the Model

I have chosen to present rates of increase in the form of finite rates over specified, discrete units of time. Example life history patterns shown here typify species that are iteroparous and long lived, that produce a single young per birth, provide extensive maternal care, and are polygynous. Definitions and model structure follow Conley (1978).

A cohort is a group of individuals of the same age in a population. A cohort of age zero individuals (that is, a group of newborn), can be followed through time, and the pattern of survival can be determined. When the cohort is exhausted, the resultant pattern is converted into a schedule of survival probabilities by:

$$l_x = n_x/n_o \qquad\qquad (1)$$

where x represents age, l_x is the probability that an age zero individual will survive to enter the xth age class, and n_x is the number of individuals in the x to x+1 age class. Survival schedules are often sex-specific; for our purposes, I will deal only with the female portion of the population under the assumptions that:

1) there are sufficient males to provide for the breeding demands of the females; and

2) that the sex-ratio at birth (expressed as proportion males) is one, and that survival probabilities are the same for males and females.

A fertility schedule for a population (m_x) typically reflects the expected births of daughter offspring for a female aged x to x+1. This includes females that are producing young, those that are not, and a term for expected clutch size. In this manner, we incorporate clutch size, a result of an ultimate evolutionary trend, and proportion of females giving birth, a more proximate ecological attribute (Conley, 1978). Thus defining F_x as the expected production (that is births) of daughter offspring for a female of age x to x+1 *that does produce* at time t, and B_x as the proportion of females age x to x+1 that are producing, the traditional m_x as defined above is given by:

$$m_x = F_x B_x . \qquad\qquad (2)$$

In the model used here, I substitute the right side of Equation 2 for m_x. In this manner, the importance of whether or not a female breeds during any given time step becomes apparent.

Thus, with l_x, F_x, and B_x as above, the finite rate of increase for a birth-pulse population at stable age distribution is given by:

$$1 = \sum_{x=1}^{\omega} \lambda^{-x} l_x F_x B_x \quad , \tag{3}$$

with ω being the last age class of reproduction, and λ the finite rate of increase.

The rate of increase, λ, may be obtained by the interactive solution of the polynomial in λ (Equation 3), or by utilizing a compatible projection model and iterating time steps through attainment of stable age distribution, and computing λ at that time.

In general, the number of young entering the population at time t is given by:

$$n_{o,t} = \sum_{x=1}^{\omega} n_{x,t} F_x B_x \quad , \tag{4}$$

and the number of individuals entering subsequent age classes at the next time step is:

$$n_{x+1,t+1} = n_{x,t} p_x \quad , \tag{5}$$

where p_x is age-specific survival given by:

$$p_x = 1 - [(l_x - l_{x+1})/l_x]$$
$$= l_{x+1}/l_x. \tag{6}$$

It should be noted that rates of increase as expressed above are compatible with classical population theory (Lotka, 1956; Hutchinson, 1978; Mertz, 1970), but are not similar to current applications of matrix projections that stem from the works of Leslie (1945, 1948), Lewis (1942), and Bernardelli (1941), where recent usage has eliminated the originally defined top row of the projection matrix (Leslie, 1945, 1948) and replaced it with a standard m_x schedule. It is beyond the scope of this paper to discuss these distinctions (see Goodman, 1967; Keyfitz, 1968; or Michod and Anderson, 1980).

Population Projections

The above model is programmed in FORTRAN IV along with various extensions not essential to this discussion. Additional examples of the use of this model are given by Conley (1978), Conley et al. (1977), Lenarz and Conley (1980), Watts and Conley (1981), Nelson (1978, 1980) and Tipton (1975).

No attempt has been made for this presentation to document extensively life history patterns of the kind discussed here. Much of the information required for an adequate analysis is simply lacking. There are, however, some aspects of feral equid biology upon which general agreement can be obtained, and I have used those data for examples.

Gestation period in the domestic horse (*Equus caballus*) is about 330 days, with considerable variation being induced by seasonal or nutritional factors (Asdell, 1964). In the domestic ass (*Equus asinus*), gestation is 365 days (Asdell, 1964). Twinning is particularly rare in *E. caballus* but does occur (Speelman *et al.*, 1944; Feist and McCullough, 1975), and is also of no demographic consequence in *E. asinus*. Copulation usually begins not sooner than two years in *E. caballus* and one year in *E. asinus*, with subsequent birth occurring in the three and two-year olds, respectively. Information on breeding proportions is scanty; Nelson (1978) reported 55 per cent of breeding age females (*E. caballus*) having foals; *E. asinus* appears to be comparable (Moelman, personal communication). Assuming one young per year and a sex ratio of one, the basic F_x value was set at 0.5 for ages 2-14. Modifications to this schedule generally involved manipulations of the proportion breeding vector except for questions involving effects of age at first breeding.

Survival schedules are also difficult to find in the formal literature. Nelson (1978, 1980) constructed a tentative l_x schedule for *E. caballus* from New Mexico, and Conley (unpublished data) reconstructed a series of age-structure patterns from various *E. caballus* populations. In general, both species appear to have high adult survival, (that is, the l_x curves are fairly flat prior to old age), and there is some suggestion that males survive less well than do females (Nelson, 1978, 1980). This may be the result of differential migration or other behavior in the males, which superficially shows as mortality in the capture data.

In order to keep the number of possible combinations of life history patterns within reason, I have chosen 14 years as maximum age; the effects of such a choice are discussed below.

Given the above general considerations, an extensive series of simulations were conducted that represent the various combinations of life history patterns of interest. Eight hypothetical survival schedules were constructed, and each was matched against sequences representing various age-at-first-breeding, proportion breeding, reproductive life spans, and maximum age patterns. All simulations were continued to attainment of stable age distribution. Subsequent finite rates of increase thus obtained represent theoretical upper boundaries; behavior of such response variables as λ during transition periods prior to stable age distribution is discussed elsewhere (Conley, Gross, and Rebar, unpublished data). All simulations were conducted on the New Mexico State University Amdahl 470-V5 computer.

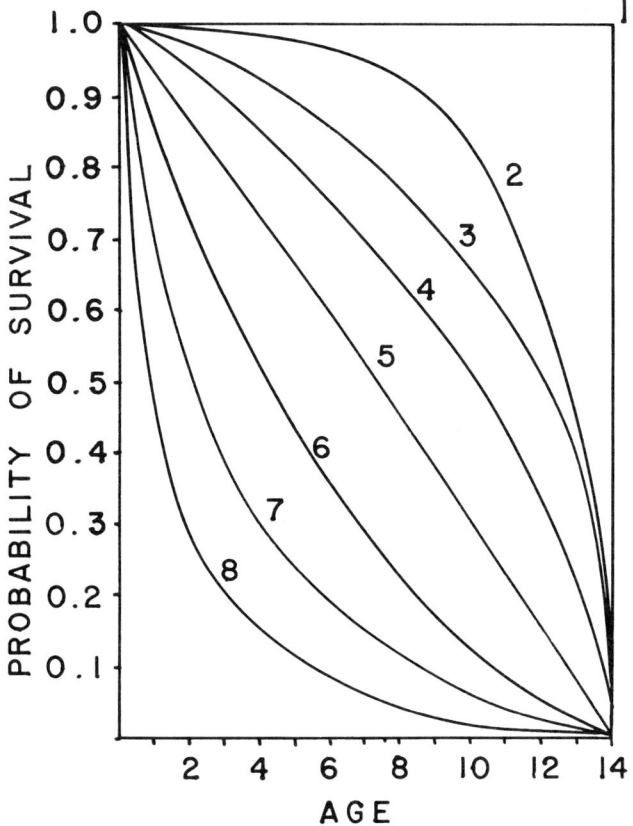

FIG. 1.—The array of survival schedules across which simulations were conducted. Schedule 1 represents a theoretical maximum; schedules 2 through 8 are simply representative of the variation available.

RESULTS

The eight hypothetical survival schedules are shown in Fig. 1. Schedule 1 represents a theoretical maximum, with no deaths prior to maximum age 14. Schedules 2 through 8 simply represent successive decreases in survival rates.

Initially, an important decision involved what age to use for the maximum. Most large-mammal populations are aged according to various tooth eruption and wear patterns. Such techniques typically provide somewhat less than satisfactory results through the stage where all teeth are at occlusal level, and increasingly notorious results in older animals. Thus, the question arises: how important is maximum attained age in determining rate of increase? The answer can be seen in Fig. 2, which suggests that knowing maximum age is not particularly important. The curves representing λ at various maximum ages inflect and tend to flatten at about age 14.

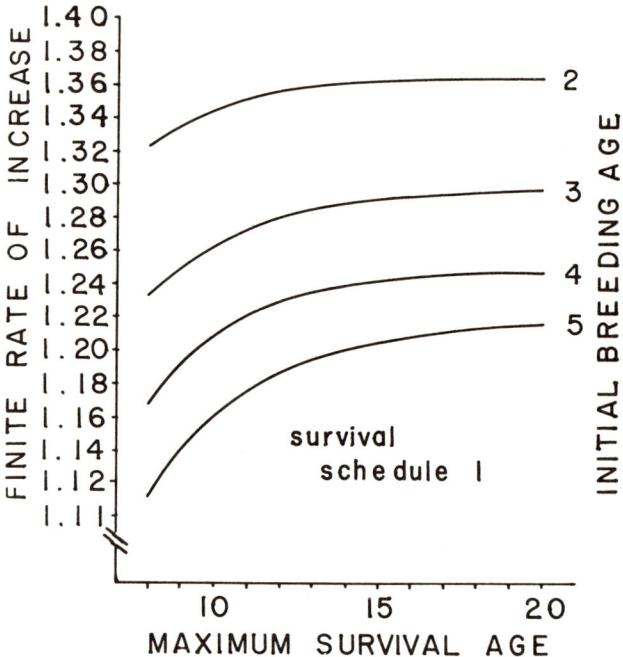

FIG. 2.—Finite rates of increase (λ) obtained at various maximum ages and at four differing ages at first breeding schedules. Note that these simulations were conducted using the maximum theoretical survival schedule, and 100 per cent breeding across all breeding-age females.

This flattening effect would be even more pronounced given realistic survival schedules, where the proportion of older individuals in the population would be considerably less than that resulting from survival Schedule 1. It is this effect that led to the decision to use 14 as "generalized-old-age;" the results described here would not be greatly different if this age were increased by several time units. The conclusion is that summarization techniques that combine older age classes for subsequent analysis are probably appropriate, and that accurate aging techniques for these categories would (even if available) simply provide a level of detail to which the rate-of-increase question is insensitive.

The question of whether postreproductive individuals are important to rates of increase has been alluded to occasionally (Allee *et al.*, 1949; Cole, 1954; Wilson, 1975; Tipton, 1975), but little substantive data seems to exist.

In contrast, age at first breeding (that is, parturition) is important in determining rate of increase. The four curves shown in Fig. 2 reflect rates of increase attained across various maximum ages, with breeding beginning in years two through five. Finite rates of increase at maximum age 14 vary from approximately 1.2 to 1.35 and represent an increase in doubling time in the population by about a factor of two. The relative positions of the

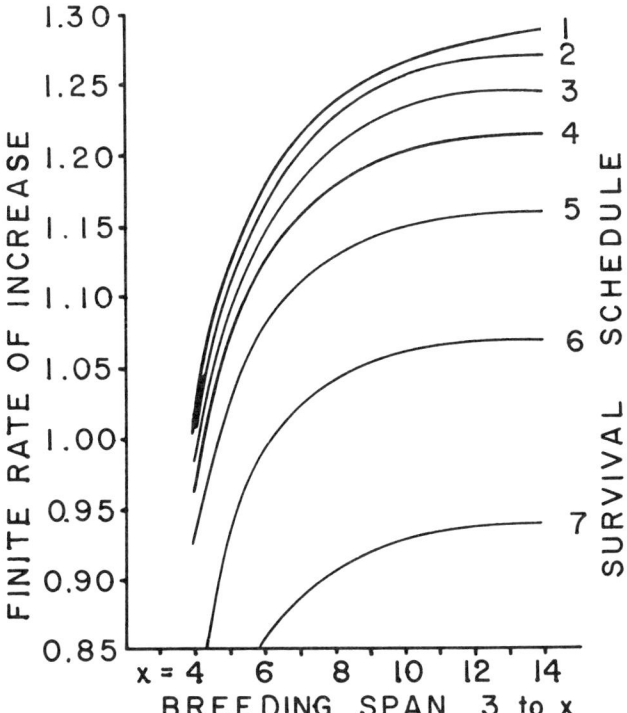

Fig. 3.—Finite rates of increase (λ) obtained at various durations of breeding across survival schedules 1 through 7. Survival schedule 8 is off the graph and thus not presented. These simulations were conducted using 100 per cent breeding across all breeding-age categories.

curves shown in Fig. 2 would remain for simulations conducted across the various survival schedules, but, of course, would be progressively lower overall. In a real population, one would expect differing proportions of the females to begin breeding at different ages, with the average age somewhere between three and five years depending on species and conditions.

In a similar vein as with the maximum age question, I have conducted a series of simulations that provide rates of increase for the eight survival schedules across varying reproductive spans (Fig. 3). Again, the curves tend to become flat at about age 10 or 11, suggesting that reproductive life span is important only when it is very brief, and progressively less important as it extends beyond six or eight years (that is, ages 9-11). Thus, although it is important to have information on breeding ages, such information has a decreasing effect on rate of increase at older ages.

Distinguishing between the expected number of female offspring to be produced by a female that is producing, and those females that are of breeding age but not producing allows considerable insight into the dynamics of population growth. The number of young a female of a given species can produce is a product of long-term evolution and involves

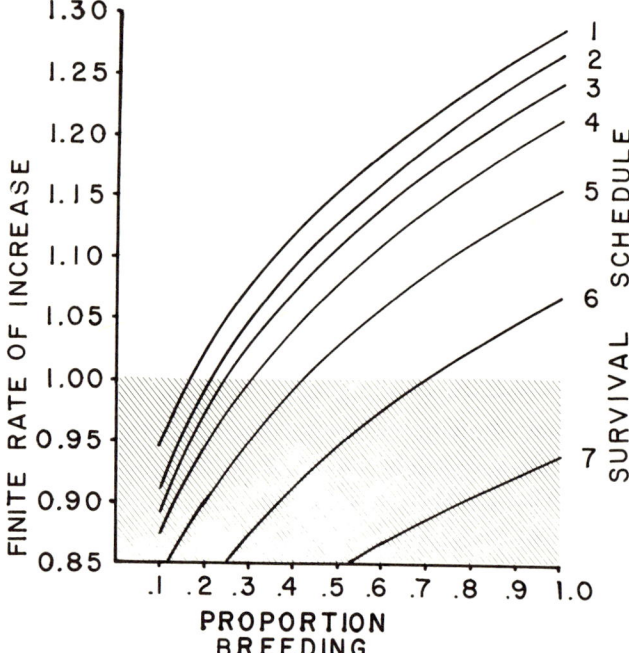

Fig. 4.—Finite rates of increase (λ) obtained at various proportions breeding across survival schedules. Age at first breeding was three for all simulations, and proportion breeding was constant across breeding ages within simulation sets.

considerations such as size of female and embryos, skeletal structure, physiological and energetic demands of the young and capabilities of the mother, extent of maternal care, and so forth. It is thus no mistake, from an evolutionary viewpoint, that both *E. caballus* and *E. asinus* typically have one foal per birth. This being so, and assuming a sex ratio (proportion males) at birth of one, F_x values are 0.5 for all age classes where breeding occurs.

In contrast to the ultimate reasons why a species has a given expectation of young at birth, the determination of whether or not a female joins the breeding component involves more proximate ecological and behavioral information. Separating these factors in the above model through the use of the B_x vector thus focuses attention on the proximate causes involved in population potentials, even though, algebraically, the results are similar to those treated in a more standard fashion.

The effects of varying proportion breeding on rates of increase for seven survival schedules are shown in Fig. 4. Those curves are quite steep, and provide evidence for the importance of determing proportion breeding in populations of *E. caballus* and *E. asinus*.

Given the various interactions of demographic variables shown in Figs. 2 through 4, and the resultant rates of increase, the question of interpretation

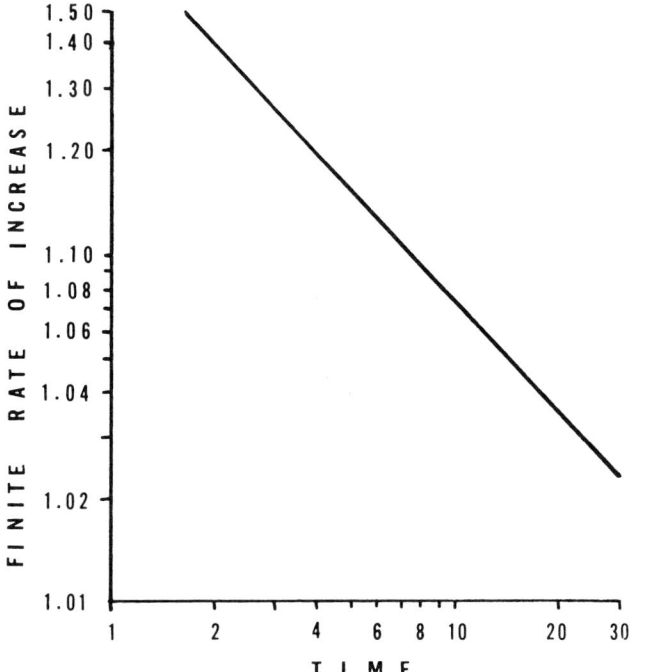

Fig. 5.—Doubling times resulting from various finite rates of increase.

becomes important. Using the fact that $\lambda = e^r$, where r is the instantaneous rate of increase, population doubling time is (Conley, 1978):

$$T_d = \frac{\ln 2}{\ln \lambda} \quad .$$ (7)

For the range of λ values representing those shown in Figs. 2 through 4, Fig. 5 shows a graph of population doubling times as a function of λ. Doubling times are quite short for $\lambda \geq 1.07$ (that is, less than 10 years). The ecological importance of these potentials is discussed below.

DISCUSSION

The sensitivity of rates of increase to changes in various population attributes and life history patterns, is a subject of considerable importance to attempts to estimate such rates from wild populations. Because the rate of increase is such a highly integrative parameter, incorporating information from all levels of the biology of a species, the final values obtained are often questionable; furthermore, assumptions required by the mathematical structure are particularly restrictive, and often, appropriate data are simply impossible to obtain.

As can be seen from Figs. 1 through 4, the rate of increase is particularly sensitive to: 1) the shape of the survival function; 2) proportion of the adult female population that actually produces young; and 3) the age at first breeding. In contrast, the rate of increase is relatively insensitive to: 1) maximum age attained by the breeding females; and 2) the presence of post-reproductive animals. The latter two attributes have a decreasing effect and, within our abilities to estimate ages and breeding span in wild populations, can be effectively ignored beyond about 14 years of age and six to eight years of reproductive life, respectively.

The survival patterns shown in Fig. 1 represent an array of possible patterns, but are probably too simple to be related directly to wild populations. It is reasonable to expect higher age-specific mortality in the zero age class (that is, probability of surviving from $x = 0$ to $x = 1$). Although such patterns could be simulated, the number of combinations would increase drastically and appropriate data for comparison are currently lacking. Thus, the patterns evaluated in this study, with their resultant rates of increase, must be considered as generally conservative. This is particularly true of Fig. 2, where only the theoretical maximum survival schedule was utilized.

The question of what values for finite rates of increase can be reasonably expected in wild equid populations can be answered in part. Assuming that female survival schedules in wild populations are approximately similar to schedule 4 in Fig. 1, and that proportion breeding is on the order of 50 or 60 per cent, finite rates of increase of about 1.05 are to be expected.

Additionally, if survival schedules in the males are lower than those of females, the results presented here are higher than would be obtained in wild populations. Again, the λ values presented here are conservative.

At stable age distribution, finite rates of increase higher than 1.20 can be obtained only (see Figs. 3, 4) if the real survival schedules are similar to schedules 1, 2, or 3 (Fig. 1), *and* if the proportion breeding is 0.8 or greater across all age classes, *and* if age at first breeding is three years, *and* if breeding span beginning at age three, extends beyond about age 8 to 10. Although adequate data currently do not exist to provide a definitive answer, the conclusion from data that are available is inescapable; empirical values for the various population attributes considered here are simply too low to conclude that rates of increase in wild equid populations approach 20 per cent, much less exceed that level. Higher rates could obtain, but only on a short term basis; such rates, resulting from unnatural sex ratios, are unstable and tend to damp out quickly (Conley, Gross, Rebar unpublished data).

The presentation of population doubling times across various finite rates of increase (Fig. 5) is designed to illustrate the point that even at rates of increase between 1.01 and 1.10 (that is 1 per cent to 10 per cent) populations with similar demographic patterns do increase, with the relative concern that must be attached depending on whether the population doubles every six or so years (10 per cent increase) or whether the doubling time approaches infinity (as $\lambda \rightarrow 1.0$).

Acknowledgments

I thank Lee Metzgar, Norm Slade, and an anonymous reviewer for critique and helpful suggestions.

Literature Cited

Allee, W. C., A. E. Emerson, O. Park, T. Park, and K. P. Schmidt. 1949. Principles of animal ecology. Saunders. Philadelphia, Pennsylvania, xii+837 pp.

Anderson, D. R. 1975. Population ecology of the mallard: V. Temporal and geographic estimates of survival, recovery, and harvest rates. USDI, Bureau of Sport Fisheries and Wildlife Resource Publ., 125:v+1-110 pp.

Asdell, S. A. 1964. Patterns of mammalian reproduction. 2nd ed. Cornell Univ. Press, Ithaca, New York, vii+160 pp.

Bernardelli, H. 1941. Population waves. J. Burma Res. Soc., 31:1-18.

Caughley, G. 1966. Mortality patterns in mammals. Ecology, 47:906-918.

———. 1967a. Parameters for seasonally breeding populations. Ecology, 48:834-839.

———. 1967b. Calculation of population mortality rate and life expectancy for thar and kangaroos from the ratio of juveniles to adults. New Zealand J. Sci., 10:578-584.

———. 1977. Analysis of vertebrate populations. Wiley-Interscience. John Wiley and Sons, New York, ix+234 pp.

Cole, L. C. 1954. The population consequences of life history pheonomena. Q. Rev. Biol., 29:103-137.

Conley, W. 1978. Population Models. Pp. 305-320, in Big Game of North America: Ecology and Management. (J. Schmidt and D. Gilbert, eds.). Stackpole Books, Harrisburg. Pennsylvania, xv+494 pp.

Conley, W., and J. D. Nichols. 1978. The use of models in small mammal population studies. Pp. 14-37, in Populations of small mammals under natural conditions. (D. P. Snyder, ed.). Spec. Publ. Ser., Pymatuning Lab. Ecol. Univ. Pittsburgh, 5:xiv+1-237.

Conley, W., J. D. Nichols, and A. R. Tipton. 1977. Reproductive strategies in desert rodents. Pp. 193-215, in Transactions of the Symposium on the Biological Resources of the Chihuahuan Desert Region, United States and Mexico. (R. A. Wauer and D. H. Riskind, eds.). Proc. Trans. Ser., National Park Serv., 3:xxii+1-658 pp.

Eberhardt, L. L. 1969. Population analysis. Pp. 457-495, in Wildife management techniques. (R. H. Giles, ed.). The Wildlife Society, Washington, D.C., vii+623 pp.

Feist, J. D., and D. R. McCullough. 1975. Reproduction in feral horses. J. Repro. Fert., Suppl., 23:13-18.

Goodman, L. A. 1967. On the reconciliation of mathematical theories of population growth. J. Royal Statis. Soc. Series A, 130:541-553.

Hutchinson, G. E. 1978. An introduction to population ecology. Yale Univ. Press, New Haven, xii+260 pp.

Keyfitz, N. 1968. Introduction to the mathematics of population. Addison-Wesley Publ. Co., Reading, Mass., xiv+450 pp.

Lenarz, M., and W. Conley. 1980. Demographic considerations in reintroduction programs of bighorn sheep (Ovis). Acta Theriol., 25(7):71-80.

Leslie, P. H. 1945. On the use of matrices in certain population mathematics. Biometrika, 33:183-212.

———. 1948. Some further notes on the use of matrices in population mathematics. Biometrika, 35:213-245.

LEWIS, E. G.   1942.   On the generation and growth of a population.   Sankya, 6:93-96.

LOTKA, A. J.   1956.   Elements of mathematical biology.   Dover Publ., New York, xxx+465 pp.

MALTHUS, T. R.   1798.   An essay on the principle of population. Reprinted by MacMillan, New York, ix+396 pp.

MERTZ, D. B.   1970.   Notes on methods used in life-history studies.   Pp. 4-17, in Readings in ecology and ecological genetics.   (J. H. Connell, D. B. Mertz, and W. W. Murdock, eds.).   Harper and Row, New York, viii+397 pp.

MICHOD, R. E., AND W. W. ANDERSON.   1980.   On calculating demographic parameters from age frequency data.   Ecology, 61:265-269.

NELSON, K. J.   1978.   On the question of male limited population growth in feral horses (Equus caballus).   Unpublished Master's Thesis, New Mexico State University, Las Cruces, New Mexico, x+68 pp.

———.   1980.   Sterilization of dominant males will not limit feral horse populations. Research paper RM-226, Rocky Mt. For. Range Exp. Sta., U.S. Forest Serv., U.S.D.A., 1-7 pp.

POSPAHALA, R. S., D. R. ANDERSON, AND C. J. HENNY.   1974.   Population ecology of the mallard: II. Breeding habitat conditions, size of the breeding populations, and production indices.   USDI, Bureau of Sport Fisheries and Wildlife Resource Publ., 115, iv+73 pp.

SLADE, N. A., AND D. F. BALPH.   1974.   Population ecology of Uinta ground squirrels.   Ecology, 55:989-1003.

SPEELMAN, J. R., W. M. DAWSON, AND R. W. PHILLIPS.   1944.   Some aspects of fertility in horses raised under western range conditions.   J. Anim. Sci., 3:223-241.

TIPTON, A. R.   1975.   A matrix structure for modeling population dynamics.   Unpublished Ph.D. Dissertation, Michigan State University, East Lansing, vi+52 pp.

WATTS, T., AND W. CONLEY.   1981.   Extinction probabilities in a remnant population of Ovis canadensis mexicana.   Acta Theriol., 26:119-131.

WILSON, E. O.   1975.   Sociobiology: The new synthesis.   Belknap Press of Harvard Univ., Cambridge, Mass., ix+697 pp.

VARIATION IN REPRODUCTION OF A SUBTROPICAL POPULATION OF PEROMYSCUS LEUCOPUS

Frank W. Judd, Gary Carpenter, and Margie Wagner

Lackey (1978) examined the usefulness of five hypotheses in explaining variation in litter size in *Peromyscus leucopus*. He concluded that none of the five was adequate and that geographic variation in reproductive parameters, particularly litter size, represented a complex and largely unpredictable phenomenon. Two reasons suggested for the unpredictability were lack of life history data for both northern and southern populations of wide ranging species (most of the data is for northern populations) and lack of estimates of the extent of seasonal and annual variation in the reproductive parameters for each population.

Peromyscus leucopus ranges from near the Canadian border south through the northern and central United States west of the Rocky Mountains to the Yucatan Peninsula. Most of the information on the species' reproductive ecology has been obtained from populations in the northern portion of its range (see reviews by Cornish and Bradshaw, 1978; Fleming and Rauscher, 1978; and Lackey, 1978). Knowledge of reproductive ecology of tropical populations is limited to information provided by Lackey (1978) for a population from Campeche, México.

Populations of *P. l. texanus* occupy a brushland habitat in the Lower Rio Grande Valley (Cameron, Hidalgo, Starr, and Willacy Counties) of Texas. The area is semiarid and subtropical and is approximately midway, latitudinally, between north temperate and tropical regions. Judd *et al.* (1978) examined the relationship between the lipid and reproductive cycles of *P. leucopus* from the Lower Rio Grande Valley based on monthly collections made during 1975. They reported that *P. leucopus* was reproductively active throughout the year and that a peak of reproductive activity occurred in May, June, and July. No information was provided on litter size or frequency.

Judd *et al.* (1978) showed that a low in fat stores coincided with the peak of reproductive activity. They suggested that lipid stores are depleted in summer because of energy costs associated with reproduction and because of declining quantity and or quality of food. In arid to semiarid habitats and in the tropics, abundance of insect populations is correlated with rainfall, as is the production of high caloric plant foods such as seeds and fruits. Thus, in those habitats, variation in fat reserves might be influenced by rainfall acting through food supply (Field, 1975), and consequently a seasonal cessation of reproduction might coincide with low abundance of high caloric foods (Cameron *et al.*, 1979). Alternatively, litter size might show marked seasonal or annual variation that is correlated with rainfall.

The variability in reproduction of a subtropical population of *Peromyscus leucopus* in wet (1975) and dry (1978) years is examined in order to 1) provide information on seasonal and annual variation in reproductive activity and litter size; 2) determine whether the pattern of reproductive activity and litter size is affected by marked difference in temperature, day length, or the pattern (or amount) of precipitation; and 3) assess the degree of intraspecific variation in selected reproductive parameters.

MATERIALS AND METHODS

Collections were made at monthly intervals from January through December in 1975 and 1978 in the Lower Rio Grande Valley of Texas. In 1975, mice were captured in live traps and break-back traps, transported to the laboratory, frozen, and sealed tightly in plastic bags within 12 hours of capture. In 1978, mice were captured in live traps and returned to the laboratory where they were lightly etherized and blood was drawn by cardiac puncture for a study of seasonal variation in hematology. The mice were then killed with an overdose of ether, sealed in plastic bags, and frozen.

Subsequently, the mice were thawed, weighed, sexed, and aged. Age categories were based on weight, tooth wear, and pelage characteristics. Mice weighing more than 16 grams, showing evidence of moderate wear on the molar cusps, and having rufous or buff pelage were considered to be adults. Mice not meeting those criteria were classed as subadults. The mice were then dissected to obtain reproductive data. Pregnant females were noted and their embryos counted and weighed. Testes were removed from males and measured to the nearest 0.05 of a millimeter with Helios dial calipers. The cauda epididymis was dissected and a smear made to check for the presence of spermatoza.

Data were analyzed on both a monthly and a seasonal basis. Winter was December, January, and February; spring, March, April, and May; summer, June, July, and August; and autumn, September, October, and November. Statistical procedures and tables were those of Sokal and Rohlf (1969) and Rohlf and Sokal (1969). A probability value less than 0.05 was considered significant.

STUDY AREA

Because much of the land in Cameron, Hidalgo, Starr, and Willacy counties is in the delta of the Rio Grande, these four counties are referred to as the Lower Rio Grande Valley. Blair (1950) considered the area a distinct biotic district (Matamoran) of the Tamaulipan Biotic Province because a number of tropical species reach the northern limits of their distribution in the area. Most of Cameron and Hidalgo counties has been converted to farms and citrus groves, but substantial portions of Starr and Willacy counties remain uncleared. Native vegetation also remains at places along the banks of the Rio Grande and its immediate floodplain. The native

TABLE 1.—*Comparison of precipitation (in centimeters) and temperature (in degrees Celsius) in 1975 and 1978, with the average for the preceding 24 years.*

Month	Precipitation			Temperature		
	1975	1978	24-year $\bar{X}$	1975	1978	24-year $\bar{X}$
January	2.67	10.21	3.07	15.8	11.7	15.2
February	2.13	1.37	2.82	17.2	13.5	17.2
March	0.13	0.07	1.85	21.9	19.9	20.1
April	0.05	1.98	3.81	24.8	24.5	24.2
May	13.11	0.38	5.51	27.4	28.1	26.5
June	8.41	1.63	7.72	27.9	29.4	28.5
July	25.35	0.30	3.71	27.7	30.5	29.4
August	15.14	3.22	4.70	28.4	30.1	29.6
September	14.27	9.37	11.00	26.0	27.7	27.7
October	7.87	11.84	8.10	23.4	23.4	24.1
November	0.00	0.81	2.56	19.6	20.7	19.5
December	2.64	2.64	2.36	15.7	16.5	16.5
Total	91.77	43.82	57.21			

vegetation remaining in the floodplain consists of forest dominated by cedar elm (*Ulmus crassifolia*) (Blair, 1950). Thorny-brush grassland is the predominant vegetation in areas distant from the floodplain.

Table 1 shows a comparison of precipitation and temperature in 1975 and 1978 with the average for the preceding 24 years (N.O.A.A. Local Climatological Data Annual Summary for McAllen, Texas). About 71 per cent of the annual precipitation falls from May through October, with approximately one-third of the total occurring in September and October. Most precipitation results from thunderstorms; a single thunderstorm often produces all the rain for a given month. Precipitation in 1975 was more than double the long term average; that is, 91 per cent of the rainfall occurred from May through October. In 1978, the distribution of rainfall was markedly different than usual; 23 per cent of the annual rainfall occurred in January, and then drought conditions prevailed from February through July.

In summer, temperatures usually range from 32.6-35.2°C during the day and from 23.5-24.7°C at night. Average monthly temperatures in winter are generally above 15°C. Mean length of the frost-free period is 330 days. Frequently, there are no freezing temperatures during a year. Mean temperatures in January and February of 1978 were markedly lower than in 1975 and lower than the 24-year average. During other months, mean temperatures were similar in the two years, and, in both years, the monthly mean temperatures were similar to the long term average temperatures.

RESULTS

Reproductive cycle.—Examining both years together, three lines of evidence indicate that reproduction probably occurs throughout the year: 1)

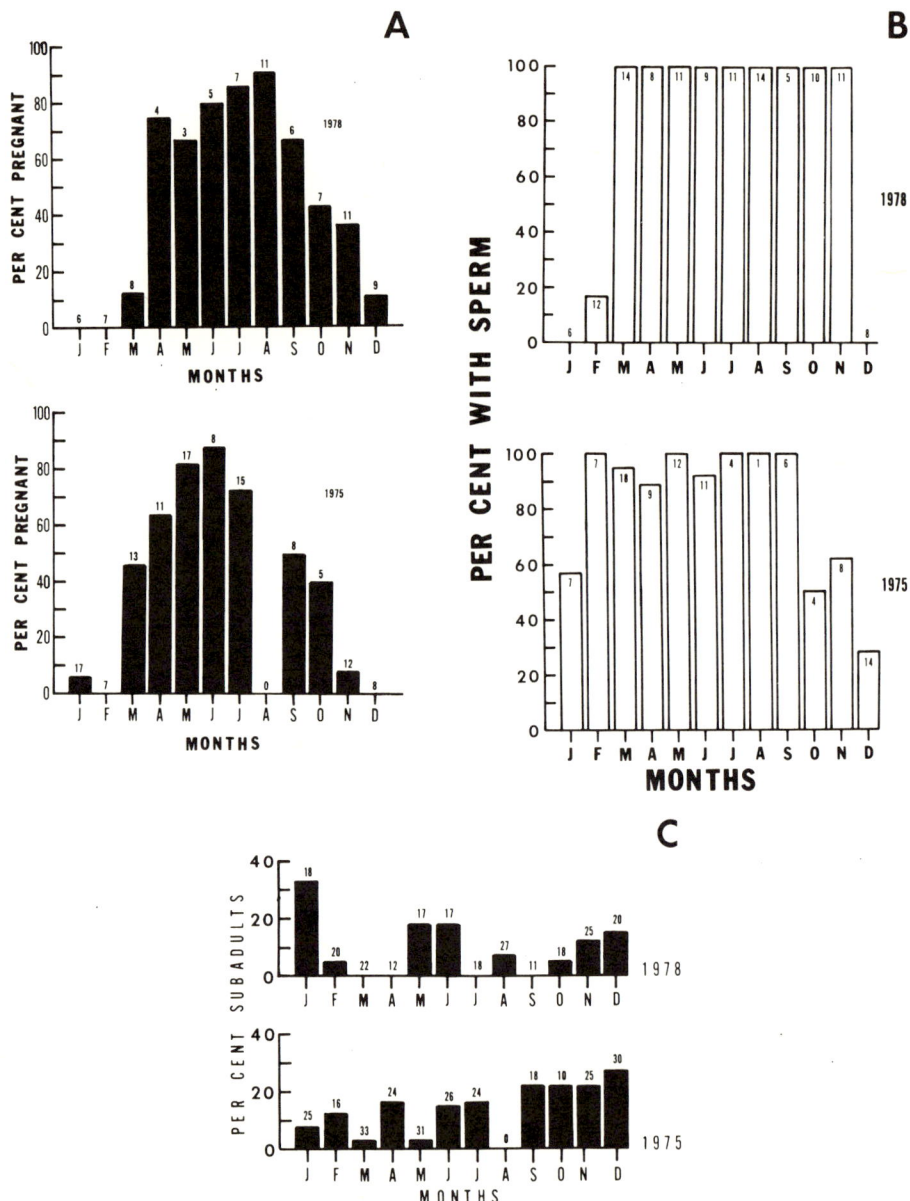

Fɪɢ. 1.—Comparison of: **A**, per cent pregnant females; **B**, per cent males with sperm; **C**, per cent subadults in the samples among months and between years. Numbers at top of bars are sample sizes.

Pregnant females were taken in every month except February (Fig. 1A); 2) Males with spermatozoa in the epididymis were present in each month (Fig. 1B); 3) Subadults were found throughout the year (Fig. 1C).

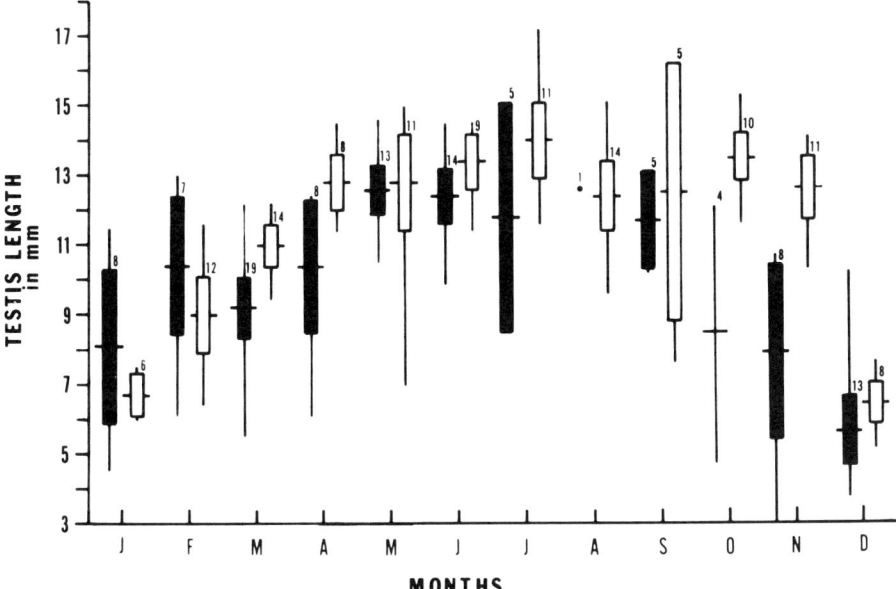

FIG. 2.—Comparison of testis length among months and between years. Vertical lines indicate the range; horizontal lines, the mean; rectangles show the 95 per cent confidence interval of the mean; 1975 is represented by closed rectangles, 1978 by open rectangles. Numbers above the rectangles indicate sample size.

The data also show that reproductive activity is highest in summer and greatly reduced in winter. When the pregnant females were grouped by season, the percentages in 1975 were: winter, 1; spring, 66; summer, 78; autumn, 28. In 1978, they were: winter, 4.5; spring, 33; summer, 87; autumn, 46.

The pattern of variation in testis length was generally similar in both years (Fig. 2). When testis length exceeded 8.0 millimeters, spermatozoa were present in the epididymes. In 1978, testis length in January and December was less than 8.0, and spermatozoa were absent. Larger testes and a higher percentage of individuals with spermatozoa were present in the October and November 1978 samples compared to samples taken in the same months in 1975.

Litter size.—Litter size ranged from two to seven in both years. Mean litter size was 3.8 in 1975 and 4.2 in 1978 (Table 2.) There was no significant difference in mean litter size between years (t=1.446, 89 df, P>0.10). The modal litter size for 1975 was four, but, in 1978, there were equal numbers of four and five.

Because the sample size in winter in both years was one individual, winter was not used in comparing litter size among seasons. There was significant variation in litter size among seasons in 1975, but not in 1978 (Table 2). In 1975, the mean for spring was significantly lower than that for

TABLE 2.—*Comparison of litter size among season in two years. Among seasons variation excludes winter.*

Season	1975			1978		
	N	$\bar{X}$	SD	N	$\bar{X}$	SD
Winter	1	2.00		1	3.00	
Spring	26	3.23	0.86	6	4.00	1.09
Summer	18	4.67	1.14	20	4.50	0.53
Autumn	6	4.67	1.37	11	3.90	0.94
Annual	53	3.83	1.35	38	4.21	1.09
	1975, F=12.102, 2, 48 df, P<0.001			1978, F=1.235, 2, 35 df, P>0.25		

summer (t=4.786, 42 df, P<0.001) and autumn (t=3.300, 30 df, P<0.01). Maximum litter sizes occurred during the period of peak reproductive activity.

Environmental correlates of reproductive activity.—Table 3 shows correlations between measures of reproductive activity and the environmental variables temperature, day length, and precipitation. Because there could be a time lag between occurrence of precipitation and its effect on vegetation, a correlation also was run between each of the measures of reproductive activity and precipitation for the preceeding month. In both years, there was

TABLE 3.—*Correlation coefficients for measures of reproductive activity and environmental variables. NS indicates probability greater than 0.05. Number of months = 12 except for per cent pregnant females in 1975 where N=11.*

Parameters	Correlation Coefficients	
	1975	1978
Temperature *vs.*% pregnant females	.958, P < .01	.972, P < .01
Temperature *vs.*% males with sperm	.643, P < .05	.822, P < .01
Temperature *vs.* $\bar{X}$ testis length	.874, P < .01	.925, P < .01
Day length *vs.*% pregnant females	.932, P < .01	.833, P < .01
Day length *vs.*% males with sperm	.711, P < .01	.696, P < .05
Day length *vs.*$\bar{X}$ testis length	.845, P < .01	.744, P < .01
Precipitation *vs.*% pregnant females	.551, NS	−.181, NS
Precipitation *vs.*% males with sperm	.548, NS	−.198, NS
Precipitation *vs.*$\bar{X}$ testis length	.657, P < .05	−.157, NS
Preceding month's precipitation *vs.*% pregnant females	.325, NS	−.317, NS
Preceding month's precipitation *vs.*% males with sperm	.261, NS	.041, NS
Preceding month's precipitation *vs.*$\bar{X}$ testis length	.546, NS	.109, NS

*Percentages arcsine tranformed.

TABLE 4.—*Comparison of length of reproductive season among populations of* Peromyscus leucopus.

Location	Duration	Authority
Ontario	March-September	Harland *et al.*, 1978
Wisconsin	March-September	Long, 1973
Michigan	March-October	Lackey, 1973
Ohio	March-October	Rintamaa *et al.*, 1976
Illinois	March-October	Hansen and Batzli, 1979
West Virginia	February-October	Cornish and Bradshaw, 1978
Texas	November-May	Bradshaw, 1962
Texas	October-June	Bradshaw, 1962
Texas	September-April	Blair and Kennerly, 1959

a significant positive correlation between day length and temperature and each of the measures of reproductive activity. There was only one significant correlation between precipitation and a measure of reproductive activity (testis length, 1975). Apparently, the pattern of precipitation does not greatly influence the annual pattern of reproductive activity.

DISCUSSION

The breeding season of *Peromyscus leucopus* populations vary geographically and temporally (Table 4). The typical north temperate cycle involves cessation of reproduction in winter. Conversely, previous reports for Texas populations (Blair and Kennerly, 1959; Bradshaw, 1962) indicate that reproductive activity is confined to the colder months of the year, with no breeding activity evident in summer. Our data differ markedly from that previously reported for Texas populations. We found that reproduction occurs throughout the year, but is low in winter. Thus, the cycle of the Lower Rio Grande Valley populations is relatively similar to that of north temperate populations. The major difference is that reproductive activity declines markedly in winter in the Lower Rio Grande Valley populations rather than ceasing completely.

Lackey (1973) did not collect in each month of the year in Campeche, but he interpreted the data as indicating that reproduction occurred throughout the year. For example, he found that 64 per cent of 22 adult females live-trapped in December and January were pregnant and produced litters. Thus, it appears that there is no hibernal cessation of reproductive activity in tropical and subtropical populations of *P. leucopus*.

Our data show a unimodal pattern of breeding. Several workers, for example, Burt (1940) in Michigan, Long (1968) in Illinois, Rintamaa *et al.* (1976) in Ohio, Svendson (1964) in in Kansas, and Brown (1964) in Missouri have reported a bimodal breeding pattern with peaks in spring and fall separated by a low or cessation in summer. However, a summer decline or cessation in reproduction might not be consistent within a given population from year to year. For example, Batzli (1977) reported that in Illinois during midsummer of 1972 and 1974, 50-60 per cent of the females were reproducing, but in 1973, reproduction ceased entirely during midsummer. Cornish and Bradshaw (1978) reported that the breeding season in West Virginia was divided into spring and summer efforts by a reduced effort in June. They attributed the apparent reduced effort in June to young of the year becoming old enough to be counted in the adult sample rather than to a real decrease in reproductive activity. Recruitment of nonbreeding young adults could explain apparent declines or short periods of apparent cessation of reproductive activity in other north temperate populations, but it does not explain the extended estival period of nonbreeding in Texas populations.

Brown (1964) noted a summer reproductive decline in one of two years for *P. leucopus* in Missouri, which he suggested might be related to a lack of free water. In our study, rainfall was very low throughout the spring and summer in 1978, but there was no decrease in reproductive activity. If the effect of reduced rainfall is a reduction in insect populations and seed crop, rainfall in the Lower Rio Grande Valley might not affect those factors because of widespread irrigation. Also, many of our specimens came from the immediate floodplain of the Rio Grande, where a subirrigation effect might be important in maintaining vegetation and insect populations.

In nonirrigated ranching areas of Texas brushlands, a combination of reduced rainfall and high temperatures for extended periods of time might cause cessation of reproduction. For example, the years 1952 to 1957, when Blair and Kennerly (1959) conducted their study, was an extended drought. Except for 1954, rainfall each year was one-half (or less) of the long term average. In several years, total rainfall was less than 25.4 centimeters. Also, temperatures were higher than usual. Because the area where their study was conducted (George West, Texas) is largely nonirrigated ranchland, the reduced rainfall might have affected the food supply. *Peromyscus* species are known to estivate under conditions of high temperatures and reduced food supply (Morhardt and Hudson, 1966); thus, these conditions might have been responsible for the cessation of reproduction in summer.

The rather symmetrical unimodal pattern of reproductive activity exhibited by populations in the Lower Rio Grande Valley suggests that proximate environmental factors cueing the reproductive cycle should vary in the same way. Both day length and temperaure do so; that is, they are low in winter, peak in summer, and vary symmetrically on either side of the peak. Therefore, it is not surprising that we found significant positive correlations between each measure of reproductive activity and day length and temperature. Whitaker (1940) showed that increasing photoperiod in spring is related to the onset of breeding. Similarly, decreasing day length in autumn is a factor in terminating breeding activity. Lynch (1973) concluded that changing photoperiod is probably responsible for both regression of the reproductive system and the autumnal molt in wild populations of *P. leucopus*. He showed that differences in photoperiod had a major effect on the reproductive system, whereas differences in temperature had little effect. Furthermore, the magnitude of the low temperatures in winter in the Lower Rio Grande Valley are similar to those that regularly occur in fall in north temperate populations such as Wisconsin and Michigan. Therefore, the most likely proximate factor cueing the reproductive cycle is photoperiod. Differences in the pattern, or totals, of precipitation did not affect the pattern of reproduction or reproductive effort (as evidenced by litter size).

One explanation for the decline in reproductive activity in winter (but not the complete cessation) might be that day length and temperature reach critical low values for most, but not all, individuals of the population. Alternatively, food availability or quality in winter might be insufficient for most individuals to support reproduction in addition to the costs of maintenance.

Millar (1978) concluded that litter size in *P. leucopus* is quite constant geographically (means of four to five). He stated that the greatest variation in litter size is seen within populations at any one time and suggested that litter size might be related primarily to the condition of the females and the attributes of their individual home ranges rather than to season or geographic area. Millar (1978) attributed some of the geographic variation in litter sizes to sampling biases, noting that overwintering females will be more abundant in samples taken early in the breeding season and in samples in populations with short breeding seasons, that is, in northern populations. Both he and Fleming and Rauscher (1978) showed that older females produce larger litters. Therefore, litter sizes in northern populations are larger because the age structure is shifted in favor of older and more fecund females. Conversely, in southern populations with long breeding seasons, the mean litter sizes may be lower because more young of the year have an opportunity to breed during the summer or autumn of their birth.

Our results support Millar's (1978) conclusion that infra-annual, intrapopulational variation in litter size is greater than geographic or annual variation. The seasonal variation we observed might be related to the age structure of the population, but this study did not provide sufficient data to resolve the question. Future research should focus on determining the frequency of litters, age-specific fecundity, and age-specific mortality to permit comparison of reproductive effort in northern and southern populations. This probably can best be accomplished with mark-recapture studies using nest boxes.

CONCLUSIONS

Peromyscus leucopus in the Lower Rio Grande Valley of Texas reproduces throughout the year, but reproductive activity is lowest in winter. There is no evidence of cessation of reproductive activity in summer; rather, the peak of reproductive activity occurs at this time. The pattern of reproduction and reproductive effort are not sensitive to variation in the annual pattern of precipitation. This might be due to the presence of an extensive network of irrigation ditches and drainage canals, and a subirrigation effect for Rio Grande floodplain communities. However, drought extended over a period of years could reduce significantly food availability and quality, which might result in reduction or cessation of reproduction in summer. The most likely proximate factor cueing the reproductive cycle is photoperiod. Mean litter size is approximately four, and seasonal variation is greater than annual variation. Seasonal variation in litter size might reflect differences in the age structure of the population.

ACKNOWLEDGMENTS

This paper is dedicated to the memory of Robert L. Packard. His knowledge and curiosity about ecology, evolution, and life histories of mammals contributed to my interests and thinking in many ways. He will be remembered with appreciation and affection by the many students whose careers he helped to mold.

This study was supported, in part, by Grant No. RR 08038 from the Division of Research Resources, National Institutes of Health, which is gratefully acknowledged. Thanks go to Jane Judd for typing the manuscript.

LITERATURE CITED

BATZLI, G. O. 1977. Population dynamics of the white-footed mouse in floodplain and upland forests. Amer. Midland Nat., 97:18-32.

BLAIR, W. F. 1950. The biotic provinces of Texas. Texas J. Sci., 2:93-117.

BLAIR, W. F., AND T. E. KENNERLY, JR. 1959. Effects of X-irradiation on a natural population of the wood-mouse (*Peromyscus leucopus*). Texas J. Sci., 11:137-149.

BRADSHAW, W. N. 1962. Variation of the wood-mouse, *Peromyscus leucopus*, under linear patterns of distribution in Texas. Unpublished Ph.D. dissertation. Univ. Texas, Austin, 153 pp.

BROWN, L. N. 1964. Reproduction of the brush mouse and white-footed mouse in the central United States. Amer. Midland Nat., 72:226-240.

BURT, W. H. 1940. Territorial behavior and populations of some small mammals in southern Michigan. Misc. Publ. Mus. Zool., Univ. Michigan, 45:1-58.

CAMERON, G. N., E. O. FLEHARTY, AND H. A. WATTS. 1979. Geographic variation in the energy content of cotton rats. J. Mamm., 60:817-820.

CORNISH, L. M., AND W. N. BRADSHAW. 1978. Patterns in twelve reproductive parameters for the white-footed mouse (*Peromyscus leucopus*). J. Mamm., 59:731-739.

FIELD, A. C. 1975. Seasonal changes in reproduction, diet and body composition of two equatorial rodents. E. African Wildlife J., 13:221-235.

FLEMING, T. H., AND R. J. RAUSCHER. 1978. On the evolution of litter size in *Peromyscus leucopus*. Evolution, 32:45-55.

HANSEN, L. P., AND G. O. BATZLI. 1979. Influence of supplemental food on local populations of *Peromyscus leucopus*. J. Mamm., 60:335-342.

HARLAND, R. M., P. J. BLANCHER, AND J. S. MILLAR. 1979. Demography of a population of *Peromyscus leucopus*. Canadian J. Zool., 57:323-328.

JUDD, F. W., J. HERRERA, AND M. WAGNER. 1978. The relationship between lipid and reproductive cycles of a subtropical population of *Peromyscus leucopus*. J. Mamm., 59:669-678.

LACKEY, J. A. 1973. Reproduction, growth, and development in high-latitude and low-latitude populations of *Peromyscus leucopus* (Rodentia). Unpublished Ph.D. dissertation, Univ. Michigan, Ann Arbor, 128 pp.

———. 1978. Reproduction, growth, and development in high-latitude and low-latitude populations of *Peromyscus leucopus* (Rodentia). J. Mamm., 59:69-83.

LONG, C. A. 1968. Populations of small mammals on railroad right-of-way in prairie of central Illinois. Trans. Illinois State Acad. Sci., 61:139-145.

———. 1973. Reproduction in the white-footed mouse at the northern limits of its geographical range. Southwestern Nat., 18:11-20.

LYNCH, G. R. 1973. Seasonal changes in the thermogenesis, organ weights, and body composition in the white-footed mouse, *Peromyscus leucopus*. Oecologia, 13:363-376.

MILLAR, J. S. 1978. Energetics of reproduction in *Peromyscus leucopus*: the cost of lactation. Ecology, 59:1055-1061.

MORHARDT, J. E., AND J. W. HUDSON. 1966. Daily torpor induced in white-footed mice (*Peromyscus* spp.) by starvation. Nature, 22:1046-1047.

RINTAMAA, D. L., P. A. MAZUR, AND S. H. VESSEY. 1976. Reproduction during two annual cycles in a population of *Peromyscus leucopus noveboracensis*. J. Mamm., 57:593-595.

ROHLF, F. J., AND R. R. SOKAL. 1969. Statistical Tables. W. H. Freeman and Co., San Francisco, xi+253 pp.

SOKAL, R. R., AND F. J. ROHLF. 1969. Biometry. W. H. Freeman and Co., San Francisco, xiii+776 pp.

SVENDSON, G. 1964. Comparative reproduction and development in two species of mice in the genus *Peromyscus*. Trans. Kansas Acad. Sci., 67:527-538.

WHITAKER, W. L. 1940. Some effects of artificial illumination on reproduction in the white-footed mouse, *Peromyscus leucopus noveboracensis*. J. Exper. Zool., 83:33-60.

REPRODUCTION IN THE SOUTHERN PLAINS WOODRAT (NEOTOMA MICROPUS) IN WESTERN TEXAS

ROBERT W. WILEY

Woodrats, genus *Neotoma*, are widely distributed in North America. The 20 recognized species occupy a variety of habitats from low, hot, dry deserts or humid jungles to rocky slopes above timberline (Hall, 1981). There is extensive literature on *Neotoma*, but papers treating reproductive parameters are few and limited in detail. Most of those accounts contain only cursory observations made while conducting other phases of systematic or natual history studies.

The most informative papers on reproduction of members of the genus are as follows: for the eastern woodrat, *N. floridana*, Poole (1940), Pearson (1952), Hamilton (1953), Rainey (1956), Spencer (1968), Knoch (1968), and Birney (1970); for the dusky-footed woodrat, *N. fuscipes*, English (1923), Donat (1933), Wood (1935), Vestal (1938), and Linsdale and Tevis (1951); for the white-throated woodrat, *N. albigula*, Feldman (1935), Vorhies and Taylor (1940), and Richardson (1943); for the southern plains woodrat, *N. micropus*, Feldman (1935), Johnson (1952), Raun (1966), Spencer (1968), and Birney (1970); for the bushy-tailed woodrat, *N. cinerea*, Egoscue (1962); for the Mexican woodrat, *N. mexicana*, Brown (1969); and for the desert woodrat, *N. lepida*, Egoscue (1957). No data have been published about reproduction of Stephens' woodrat, *N. stephensi*, and reproductive data for the species inhabiting México are practically nonexistent.

Comprehensive studies on the reproductive biology of *N. micropus* have been few. Spencer (1968) and Birney (1970) have investigated hybridization between *N. micropus* and *N. floridana* whereas Anderson (1969) has done likewise for *N. micropus* and *N. albigula*. Notes on reproduction of *N. micropus* have been recorded by Bailey (1931), Feldman (1935), Blair (1943), and Holdenried and Morlan (1956), for New Mexico, Warren (1926) and Finley (1958) for Colorado, Hall (1955) for Kansas, Spencer (1968) and Martin and Preston (1970) for Oklahoma, Baker (1956) for Coahuila, and Dice (1937) and Alvarez (1963) for Tamaulipas.

Although the southern plains woodrat is one of the most widespread of the small mammals inhabiting the brushlands of western Texas, information on reproduction of this species within the State is scant. Davis (1974) reported briefly on reproduction. Life history studies conducted in the brushlands of southern Texas by Johnson (1952) and Raun (1966) contain brief sections on reproduction. No extensive reproductive study on *N. micropus* has ever been attempted on a year-around basis. The objective of this study was to determine the annual reproductive dynamics of *N. micropus* near the geographical center of its range in western Texas.

METHODS AND MATERIALS

Woodrats were collected at monthly intervals for 12 consecutive months, from October 1970 through September 1971. All animals were collected approximately one mile south and two miles east of Wink, Winkler County, Texas, at an elevation of 2800 feet. This region lies in the ecotone between the Kansan and Chihuahuan Biotic provinces (see Blair, 1950). The dominant vegetation consists of mesquite, *Prosopis glandulosa*, catclaw acacia, *Acacia greggii*, sand sage, *Artemisia filifolia*, broom snakeweed, *Xanthocephalum sarothrae*, and soapweed, *Yucca* sp., four-wing salt bush, *Atriplex canescens*, and shin oak, *Quercus havardii*, are locally common. Grasses and forbs are present but sparse. Most woodrat houses are located at the bases of mesquite with the remainder in similar locations in catclaw acacia.

The climatology of the area is characterized by relatively mild winters and hot, dry summers. Mean January and July temperatures are 8° and 28°C, respectively, and the average growing season is 218 days. Precipitation averages between 25.4 and 38.1 centimeters yearly (Tinkle *et al.*, 1962). Weather records obtained by the author from the Winkler County Airport, 6 miles northwest of the study area, show the mean precipitation over the seven years previous to 1972 to have been 25.4 centimeters. Temperature during the period of study ranged from −17°C in January to 41°C in July. Rainfall over the same 12-month period was 28.5 centimeters.

Sandy soils are prevalent in the area, and soil temperature taken at a depth of 15.2 centimeters near the study area by Garner (1974) revealed an average winter minimum temperature of 2°C and an average maximum summer temperature of 33°C. Prevailing winds from the SSE averaged 10.3 mph for a 13-year period prior to 1966 at Midland, Texas (65 mi. ENE of the study area).

All animals were taken in live-traps as described elsewhere by Wiley (1971). From 26 to 65 woodrats were collected monthly during the period of study. All specimens were taken during the last week of each month. Of those, 10 males and 10 to 15 females were killed within 12 hours after capture and necropsied each month. The remaining animals were weighed, external reproductive organs examined, and released. In the months of breeding activity, females otherwise normally released were returned to the laboratory to obtain wild-conceived litters for study of postnatal development, growth, and litter size. In addition, those animals were used to determine the percentage of females pregnant each month.

Animals killed for necropsy were weighed, standard external measurements taken in millimeters (total, tail, hind foot, and ear lengths), and examined externally for reproductive condition (that is, testicular posititon, presence or absence of a perforate vaginal orifice, and condition of mammae). The testes, epididymides, and seminal vesicles were removed from the males. The contents of the caudal epididymis were extruded into a drop of Locke's solution and examined under a compound microscope for the presence of motile sperm. The length and width of the testes were measured to the nearest millimeter. The testes and seminal vesicles were weighed to the nearest one-hundredth gram. All testes were preserved in Bouin's fixative. The reproductive tracts of females were removed, examined for embryos, and preserved in AFA. To maintain a homogeneous sample and to prevent confusing reproductive changes resulting from maturational development, only woodrats weighing more than 220 grams were considered sexually mature (Wiley, 1972).

Laboratory examination of reproductive tracts followed the methods of Brown (1964, 1966, 1969). Ovaries were sectioned at 10 microns and mounted as interrupted serials (3-4 of every 15). These were stained in Delafield's hemotoxylin and eosin for histological examination. Corpora lutea counts of pregnant females were made and barring anovuly and polyovuly were assumed to represent the potential litter size. Visible embryos were counted and compared with corpora lutea counts to reveal the extent of preimplantation and early postimplantation embryo loss. Such comparisons also proved useful in determining the amount of transuterine migration. Resorbing embryos were noted, but not included in the embryo counts for determining litter size.

Female reproductive tracts were cleared in methyl salicylate (Brown, 1964) to count accurately the number of placental scars. Although all placental scars were counted, only the most recent scars were used to determine litter size. Standard deviations of the means were calculated for counts of embryos, placental scars, corpora lutea, and young in determining litter size.

Testes and epididymides of representative animals were sectioned and stained using procedures identical to those performed on the ovaries. These were inspected histologically for the presence or absence of sperm.

REPRODUCTION

Season of Breeding

Females

A total of 435 females was collected, of which 160 were killed and checked for embryos. Another 111 were placed in the laboratory to obtain litters previously conceived. The length of the breeding season, as determined by the presence of pregnant animals, extended from March through September (Table 1). The percentage of pregnant animals varied from month to month and probably was a conservative estimate of reproductive output. There were two distinct periods of reproductive activity (Fig. 1). In April, 42 per cent of the sexually mature individuals were pregnant followed by only 18 percent in May. Thereafter, the percentage increased through the summer to a peak of 48 per cent in August and then declined to 33 per cent in September, the last month in which pregnant individuals were collected.

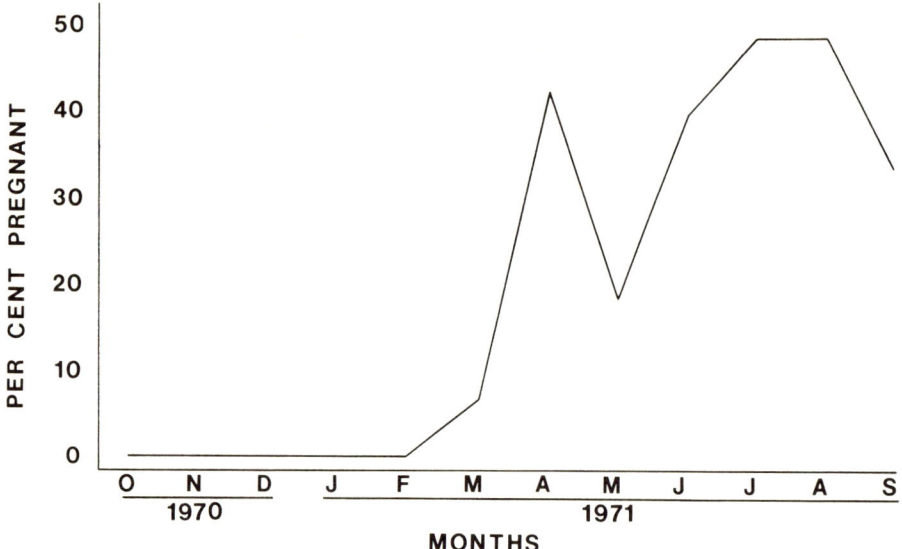

Fig. 1.—Percentage of adult females (weighing more than 220 grams) pregnant each month of the year from October 1970 through September 1971. Sample size for each month displayed above.

However, at termination of the yearlong sampling period in September of 1971, 33 per cent of the adult females were pregnant. An additional overlapping sample (not included in Table 1) of 14 females was collected and necropsied in October 1971. None was pregnant, indicating the breeding season ended rather abruptly in September or early October of 1971. Although two peaks of breeding activity were present, the latter half of the reproductive season was more productive (42 per cent of all adult females were pregnant *vs.* 27 per cent in the first half).

The season of breeding begins and ends abruptly. Only one pregnant female was collected during March and she had two near-term fetuses (crown-rump lengths of 42 millimeters and weighing 4.63 and 5.25 grams, respectively). Based on an average gestation period of 35.1 days (Spencer, 1968) and the size of the fetuses, the date of conception was likely within the first week of March. Evidence of first breeding activity in all other individuals was not observed until the last of April when 42 per cent of all adult females sampled were pregnant. Of the 15 females necropsied, seven contained embryos or small fetuses; the largest measured 19 millimeters in crown-rump length and weighed 0.93 grams. Of the 32 females placed in the laboratory, 13 bore litters. The first litter was born 9 May 1971. Although a few individuals may breed in March, most of the population begins

TABLE 1.—*Total number of females captured from October 1970 through September 1971.* .

| | 1970 | | | 1971 | | | | | | | | | |
	O	N	D	J	F	M	A	M	J	J	A	S	Total
Total number of females captured	26	26	44	39	30	38	51	34	31	38	40	38	435
Number of females released*	12	12	34	29	11	23	4	1	3	9	9	17	164
Number of females killed	10	10	10	10	15	15	15	15	15	15	15	15	160
Number sacrificed pregnant	0	0	0	0	0	1	7	2	6	7	6	6	35
Number females over 220 grams returned to lab	4	4	0	0	4	0	32	18	13	14	16	6	111
Number returned to lab pregnant	0	0			0		13	4	5	7	9	1	39**
Total number of females checked for pregnancy	14	14	10	10	19	15	47	33	28	29	31	21	271
Total number of females pregnant	0	0	0	0	0	1	20	6	11	14	15	7	74

*Includes those released, several under 220 grams returned to laboratory, and 20 that died in traps during cold weather in December of 1970.

**Includes one April female containing two embryos which died enroute to lab.

breeding the first weeks in April. The first young of the year, captured on 29 May 1971 was a 143-gram male and was undergoing the postjuvenile molt.

In September, when 33 per cent of all adult females collected were pregnant, most were nearing parturition. One of six females brought to the laboratory gave birth on 11 October 1971.

The highest percentage of females pregnant for any one month never exceeded 50 per cent of the adult female population. Inasmuch as 42 percent of the adult female population were pregnant in April, females appear synchronous in the initiation of the breeding season. The low percentage of pregnant females in May (18 per cent) is puzzling. Apparently, most sexually active females initiated breeding the first few weeks of April, and having had young by the latter part of May remain nonpregnant. As those animals breed for a second time and additional females become reproductively active, mating dates could become staggered with the percentage of animals pregnant each month increasing prior to the autumnal decline.

The condition of the nipples and development of the milk glands were accurate indices of the length of the breeding season. The nipples of all adult females examined were assigned to one of three categories; nonlactating, lactating, or postlactating. A detailed description of the reproductive changes the nipples and milk glands undergo was given by Wiley (1972).

Lactating and postlactating individuals first were taken the last week in May (Table 2). The occurrence of lactating animals at that time was expected, inasmuch as adult females were found pregnant one month

TABLE 2.—*Nipple condition of adult females (over 220 grams) in relation to lactation captured from October 1970 through September 1971.*

	1970			1971									Total
	O	N	D	J	F	M	A	M	J	J	A	S	
Females with nonlactating nipples	7	13	21	32	25	20	47	17	15	8	12	6	223
Females with lactating nipples	1	0	0	0	0	0	0	12	5	10	14	11	53
Females with postlactating nipples	2	6	1	0	0	0	0	4	9	11	5	5	42

earlier. The four postlactating animals collected extended the initiation of the breeding season back into March (back-dating four weeks for lactation, and 35 days for gestation, placed the date of conception within the last two weeks of March). However, if litters were lost, lactation might have been terminated prematurely. Differentiation of the lactating-postlactating boundary is somewhat arbitrary, and it is possible that these animals were in the latter stages of lactation. The last lactating animal taken in 1970 was captured in October. Nipple conditions indicative of postlactation were noted until December, 10 to 13 weeks after the cessation of lactation.

The condition of the vaginal orifice (closed or open) has been extensively used as a criterion for assessing reproductive activity, but, in the present study, it was highly unreliable. An approximately equal number of woodrats with the vaginal orifice open were found in all months of the year. Additionally, some females in all stages of pregnancy possessed a perforate vagina. As a result of examining a large number of females in the field and laboratory, some nonperforate females can be made to appear perforate by merely bending the tail dorsally or gently rubbing the region.

In many adult females, a membranelike structure forms either entirely or partially over the vaginal orifice. Unlike vaginal closure only (see above), the presence of the membrane indicates the season of nonbreeding. From November through March inclusively, over 30 per cent of all adult females captured possessed such a membrane. The percentage for February was 44 per cent. During the remaining months when breeding occurred, only five animals were observed with a vaginal membrane. During the study, 49 per cent of all sexually immature females (weighing less than 220 grams) captured had the vaginal opening closed by a membrane.

Although the percentage of females pregnant each month is an accurate appraisal of the percentage of the population that is breeding at any one time, it does not indicate what percentage of the female population breeds each year. The latter percentage was computed by combining the numbers of pregnant, lactating, and postlactating animals for each month during the breeding season (Table 3). For March and April the percentages were 6.7 and 42.5 per cent, respectively, and represent only animals that were pregnant. Percentage increased to a peak of 93.5 in August. The data for the months of July, August, and September suggest that more than 90 per cent

TABLE 3.—*Numbers and percentages of adult animals either pregnant, lactating, or post lactating during the months of active breeding, March through September 1971.*

	Mar.*	Apr.*	May	Jun.	Jul.	Aug.	Sep.	Totals
Sample size of females	15	47	33	28	29	31	21	203
Total number of animals pregnant, lactating, or postlactating	1	20	22	21	27	29	19	137
Per cent pregnant, lactating, or postlactating	6.7	42.5	66.6	75.0	93.1	93.5	90.4	67.4

*March and April samples are composed of pregnant animals only.

of the adult females partake in breeding by the end of the reproductive season. These percentages for the latter half of the breeding season are conservative estimates inasmuch as 1) some of the animals checked and released might have been pregnant and went unnoticed, and 2) other individuals might have had early litters, with the nipples returning to the nonlactating condition by the time of their capture. It seems plausible that all adult females breed at least once during the year. The only exceptions might be diseased animals or estrous females that failed to encounter a male. Inasmuch as a female likely would not fail to encounter a male during every estrus, she probably would not be prevented from breeding on a seasonal basis.

Males

A total of 221 males was collected, of which 120 (10 each month) were necropsied. Within a few minutes after death, the testes, epididymides, and seminal vesicles were removed for examination. Abundant motile sperm were found present in the caudal epididymides of woodrats taken during every month of the year. With few exceptions, the caudal epididymides were white in color, with prominent convoluted tubules. Upon clipping the ends off the epididymides and applying light pressure, a milky liquid was extruded that proved to be a reliable indicator of the presence of motile sperm. These observations were later verified microscopically. Most sperm smears were composed of a myriad of highly motile sperm, which filled the 100X field of the microscope.

Only three adults examined lacked sperm, two in November and one in June. They had small caudal epididymides, which were translucent, no convoluted tubules visible externally, and small lightweight testes. The combined testes weights for each individual were 0.45, 0.28, and 0.98 grams, respectively. One woodrat taken in July and another in September each had one small testis, with convoluted tubules which were not visible at the surface of the corresponding epididymis, and lacked sperm; however, the opposite sides were normal. The testicular weights for those two animals were 0.55 and 1.69 grams and 0.28 and 1.22 grams, respectively.

Although epididymal sperm smears establish the presence of sperm throughout the year in those storage structures, they only suggest, but give no direct proof, of presence of sperm within the testes. The possibility exists that in some instances the testes could have ceased production of sperm and only stored epididymal sperm remained. Representative testes were sectioned and upon examination revealed numerous sperm within the lumina of the seminiferous tubules. Figs. 2-5 demonstrate presence and quantity of testicular and epididymal sperm during two contrasting portions of the year: January, the coldest month of the year, and one in which breeding was not observed; and August, one of the hottest months of the year and one during the peak of breeding activity. Additional evidence of year-around occurrence of motile sperm was provided by adult males necropsied in the laboratory. They had abundant sperm present regardless of the month killed.

An analysis of the combined average testis weights for the period of study demonstrated no large difference between monthly weights (Fig. 6). A slight trend toward testes of lighter weights (and smaller size) during the months of October, November, December, and January was detected. Brown (1969) reported large seasonal weight change in the testes of N. mexicana, which correlates well with the season of breeding for the females.

Seminal vesicle weights were found to undergo a drastic annual cyclic weight change (Fig. 7). A marked increase in weight occurs between March and April and is followed by a corresponding decrease in weight between September and October. Those changes coincide with the initiation and termination of the breeding season of the females. During June, July, and August, seminal vesicles attained their greatest weights, and those were months in which the highest percentage of pregnant animals was taken. Seminal vesicles were found to be small (5-10 mm. in length) in October and remained so through the end of March. Beginning in late April and continuing through late September, the seminal vesicles had become much enlarged and in most cases filled the major portion of the lower abdominal cavity.

The scrotal condition of each male was recorded upon capture and utilized as an additional indicator of breeding. Experience in handling woodrats has shown that scrotal animals possess the ability to retract quickly the testes through the inguinal canal from the scrotum to the abdominal cavity. Hence a scrotal animal on being handled could retract the testes and appear nonscrotal. The testes of such an animal (when made to "strain" by rubbing a leather glove in its face, or placed on a flat surface at eye level) generally will become scrotal once again. However, a truly nonscrotal animal never demonstrated the scrotal condition under any circumstances. Upon relaxation at death all animals become scrotal. Raun (1966) has recorded the abililty of N. micropus to pass readily the testes through the inguinal canal. He noted that animals with scrotal testes one night could have nonscrotal testes the following night and vice-versa.

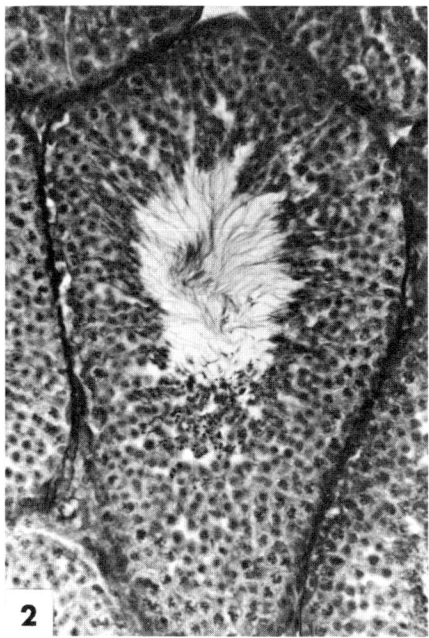

FIG. 2.—Testicular sperm in the seminiferous tubules of a male captured in January.

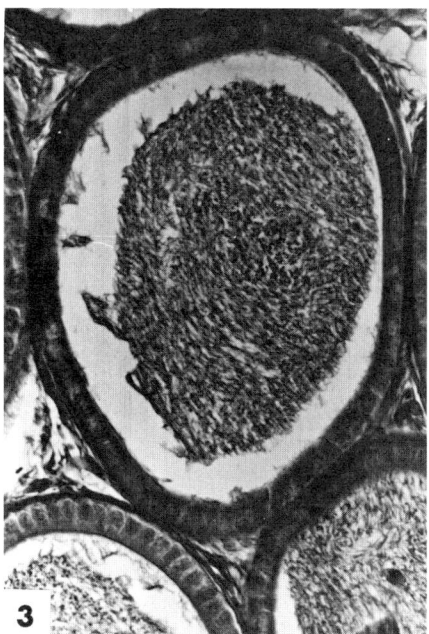

FIG. 3.—Epididymal sperm in the convoluted tubules of a male captured in January.

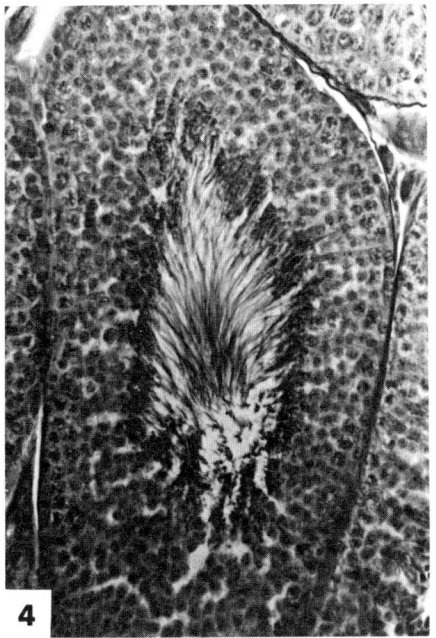

FIG. 4.—Testicular sperm in the seminiferous tubules of a male captured in August.

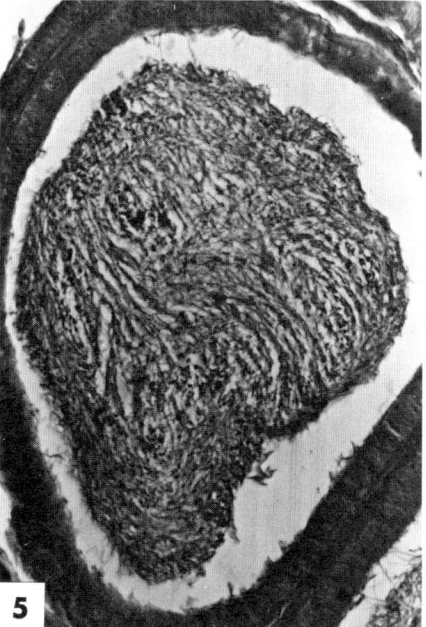

FIG. 5.—Epididymal sperm in the convoluted tubules of a male captured in August.

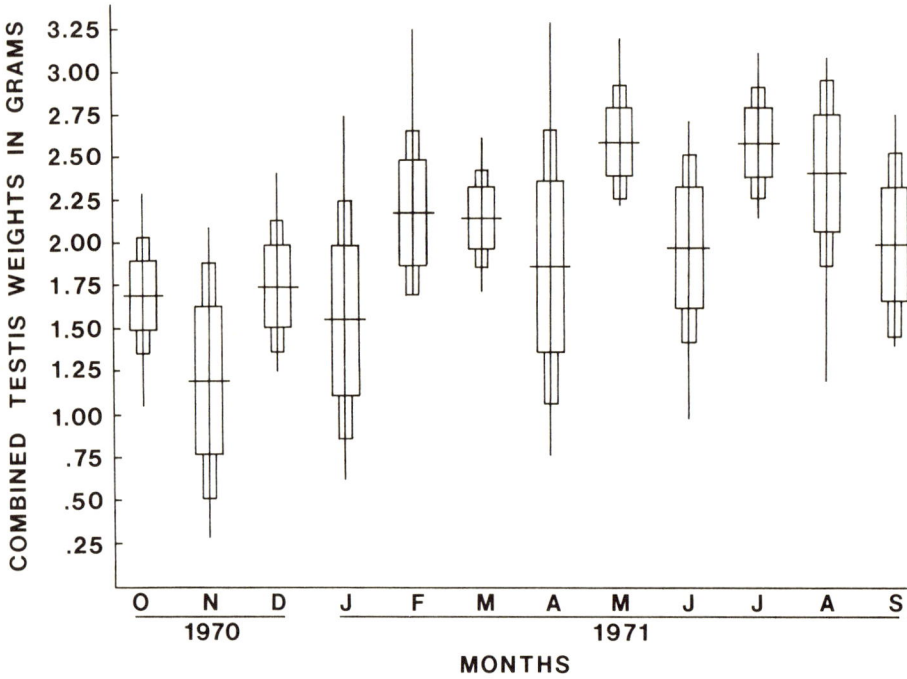

FIG. 6.—Combined testis weights by month of adult males (over 220 grams) captured from October 1970 through September 1971. The vertical line represents the range; the horizontal midline, the mean; the horizontal lines closest to the mean enclose plus or minus two standard errors of the mean; and the horizontal lines farthest from the mean enclose plus or minus one standard deviation of the mean.

Animals with scrotal testes occurred in all months of the year. However, the months that the majority of the males have scrotal testes correspond to those months in which females are pregnant (Fig. 8). Raun (1966) made similar observations. A marked change from nonscrotal to scrotal testicular position occurs between February and March, with the reverse situation occurring between September and October. All males collected during July and August were scrotal. Spencer (1968) stated that the testicular position usually is scrotal or inguinal from February to November and seemingly abdominal during December and Janaury. Of the 35 males captured weighing less than 220 grams (hence sexually immature), only three were found to be scrotal. Those individuals had weights of 195, 209, and 211 grams.

Discussion

The season of breeding for *N. micropus* began and terminated abruptly and extended from the first of March to mid-October. Johnson (1952), working in Zavala County, Texas, presented similar data on the length of the breeding season for this species. Of 244 specimens examined, he found

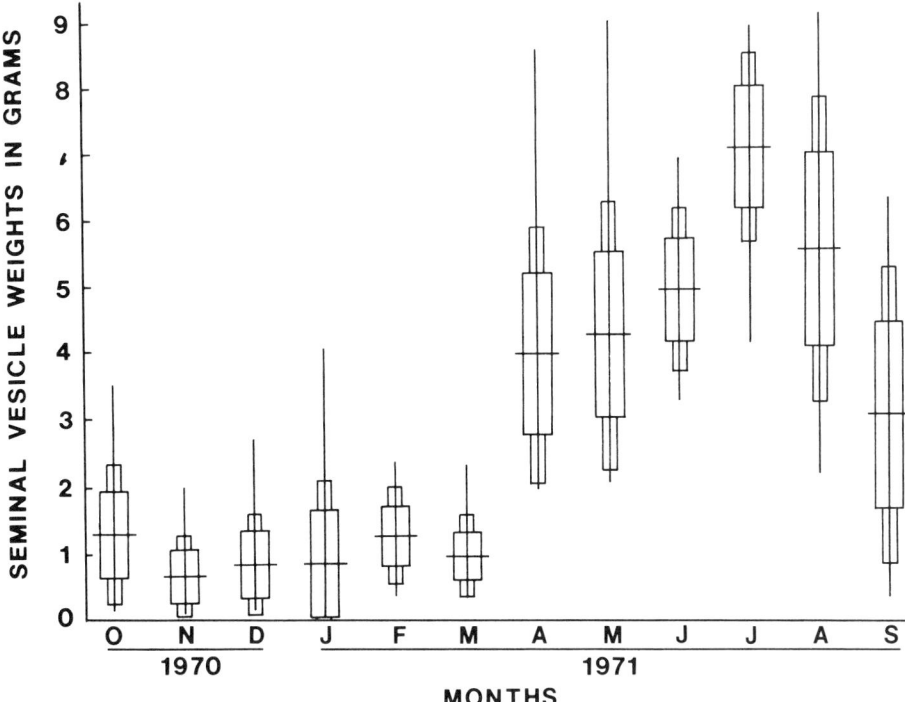

FIG. 7.—Seminal vesicle weights by month for adult males (weighing more than 220 grams) captured from October 1970 through September 1971. The vertical line represents the range; the horizontal midline, the mean; the horizontal lines closest to the mean enclose plus or minus two standard errors of the mean; and the horizontal lines farthest from the mean enclose plus or minus one standard deviation of the mean.

no pregnant females during October through January, and indicated the breeding season extended from late January through September. Overall, 17 per cent (42 of 244) of all females examined by Johnson (1952) were pregnant, whereas in my study the percentage was 27 (74 of 271).

Raun (1966) stated that no distinct breeding season exists in southern Texas (San Patricio County) and some litters are produced in all months of the year. That finding is interesting because Raun's study area was approximately 160 miles ESE of Johnson's (1952) area. Raun estimated the numbers of litters produced per month based on the number of lactating, postlactating, and pregnant females live-trapped. From those estimates, peak litter production was assigned to early spring (February, March, and April) and late autumn (November and December). The low point in breeding activity was assumed to occur in late July and August.

The apparent discrepancy concerning seasons of breeding in Texas for *N. micropus* is as follows: in my study and that of Johnson (1952), definite yearly breeding seasons are indicated, whereas, in that of Raun (1966), there is none. Several factors could explain the lack of agreement. The annual

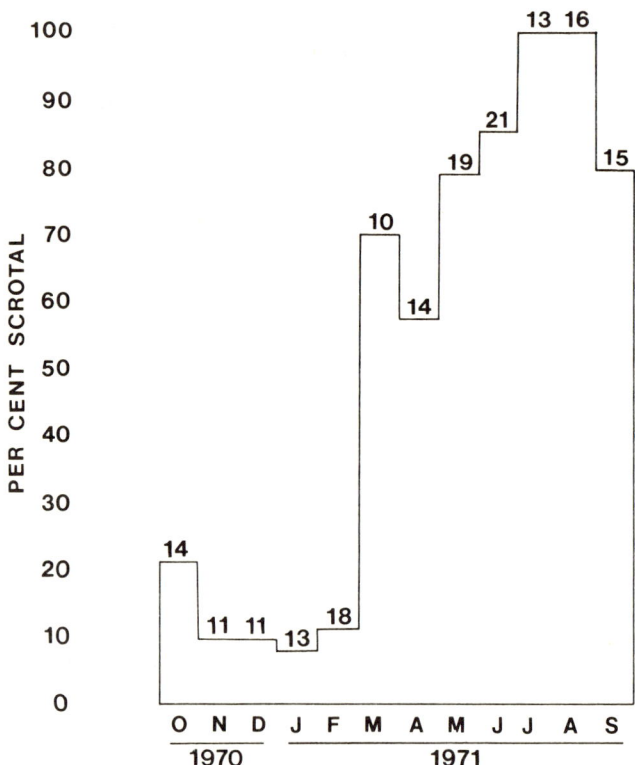

FIG. 8.—Percentage of adult males (weighing more than 220 grams) in scrotal condition from October 1970 through September 1971. Monthly sample size at top of each bar of histogram. The total sample size was 175 animals.

reproductive cycle might be modified on a year-to-year basis by climatological factors (primarily temperature and precipitation) and by population densities; the former, in most cases, influencing the latter. Raun's (1966) data suggest that the ratio of breeding to nonbreeding females increases as the population density decreases.

Weather records show the year of my study to be of average temperature and precipitation. Population density was high with 75 per cent of all houses occupied. Calculated density equaled 12 woodrats per acre in good habitat. Density of the population for several years previous to the study also was high. Johnson (1952) calculated a mean density of 20.4 woodrats per acre on his study area. Population density of Raun's (1966) study area reached a maximum of 12.6 woodrats per acre but declined to zero—primarily as a result of adverse weather, which destroyed most of their habitat. Perhaps during population lows, the length of the breeding season increases and at times might encompass the entire year.

A literature survey of *Neotoma* reveals, with few exceptions, that almost all populations of the various species studied demonstrate the existence of definite breeding seasons (for example, see Birney, 1970; Goertz, 1970; Fitch and Rainey, 1956; Spencer, 1968). Breeding generally begins in late winter to early spring and continues to late summer or late autumn with intermediate quiescent periods in some instances. Little breeding takes place in the winter months of November, December, and January. The few known exceptions are all for populations inhabiting southern latitudes.

Neotoma floridana apparently breeds throughout the year in Florida (Pearson, 1952; Hamilton, 1953) and coastal Georgia (Golley, 1962). However, the Key Largo woodrat, N. f. smalli, which is the southernmost subspecies of the eastern woodrat, demonstrated a peak in breeding period during the summer months, although limited reproduction occurred year long (Hersh, 1981). A female N. micropus with three embryos was trapped at Ft. Clark, Texas, on 13 January (Mearns, 1907). Baker (1956) listed a N. micropus from Coahuila captured in December with two embryos. In this study, under laboratory conditions, a female bred in late November gave birth to a litter on 27 December 1970. An additional mating, which occurred in late January or early February, produced a litter in March. From the foregoing, it is evident that year-around breeding can occur at least periodically and possibly regularly at some southern latitudes.

During the months when no pregnant females were captured (October through March with one pregnant animal taken in the March sample), the ovarian histology of five females from each month indicated that many contain numerous small luteinized Graafian follicles (Fig. 9). Thus, they would not have been capable of breeding, inasmuch as no ova were released. Others contained active corpora lutea of ovulation (Fig. 10). Those indicate, from a physical standpoint, capability of breeding. The presence of at least some females with perforate vaginae during all months of the year might lend additional evidence to the capability of year-around breeding.

A puzzling situation arises in that most males (from year-around presence of sperm) and at least some females (from winter presence of corpora lutea of ovulation), are capable of breeding during the months when no breeding was observed. This suggests nutritional or ethological mechanisms might curtail the breeding season at least part of the year.

The effects of nutrition on reproduction were not studied. However, the breeding season of N. micropus appeared to coincide with the major phenological changes of the vegetation. Two of the main food items utilized, mesquite and catclaw acacia, leafed out in April and lost most of their leaves by November. Leaves and fruits of those plants were the staple foods of woodrats during the breeding season; the cambium layer beneath the bark was most eaten during the nonbreeding season.

The indication of behavioral control over a portion of the reproductive season might in part be substantiated by the seasonal weight change of the testes and seminal vesicles (Figs. 6 and 7). As mentioned previously, mean testis weight remained relatively constant, but tended to be somewhat lower during the months of nonbreeding. Seminal vesicle weights were low during

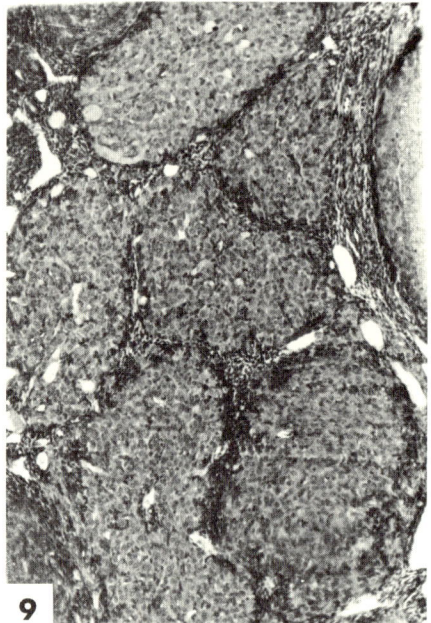

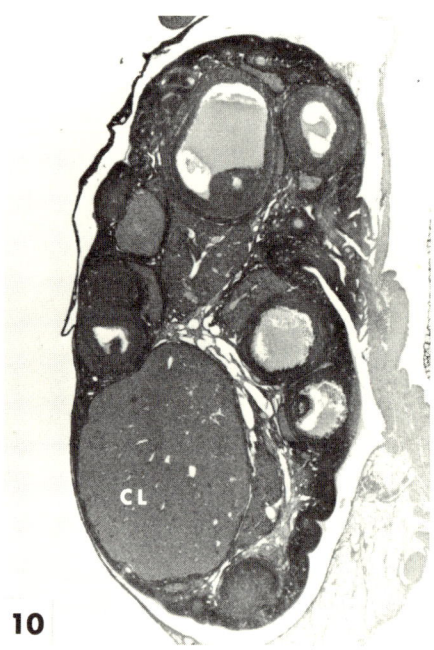

FIG. 9.—Numerous luteinized Graafian follicles from a specimen captured during a month of nonbreeding (March).

FIG. 10.—Active corpora lutea of ovulation (CL) from an individual captured during a month of nonbreeding (January).

the nonbreeding period but increased greatly during the season of breeding. The development of the accessory sex glands, including the seminal vesicles, is controlled by androgenic hormones, primarily testosterone (Short, 1972). However, size of the testes is not controlled by the testosterone level, as Nalbandov (1976) stated that testicular regression and recrudescence result from fluctuating levels of gonadotrophic hormones. Testosterone also is responsible for producing the sex drive necessary for active and aggressive breeding (Nalbandov, 1976). Therefore, it seems plausible that a large drop in seminal vesicle weights during the nonbreeding months reflects low testosterone levels. Low testosterone levels could in turn suppress the sex drive to the point of nonbreeding.

Litter Size and Prenatal Losses

Determination of litter size for *N. micropus* was calculated by counting the number of 1) young born in the laboratory to females impregnated in the field, 2) embryos in the reproductive tracts of necropsied females and 3) placental scars.

A total of 104 young was produced from 38 litters born in the laboratory, from 9 May to 11 October. Litter size ranged from one to four, with a mean of 2.73±0.83. Thirty-five pregnant females were necropsied from March to

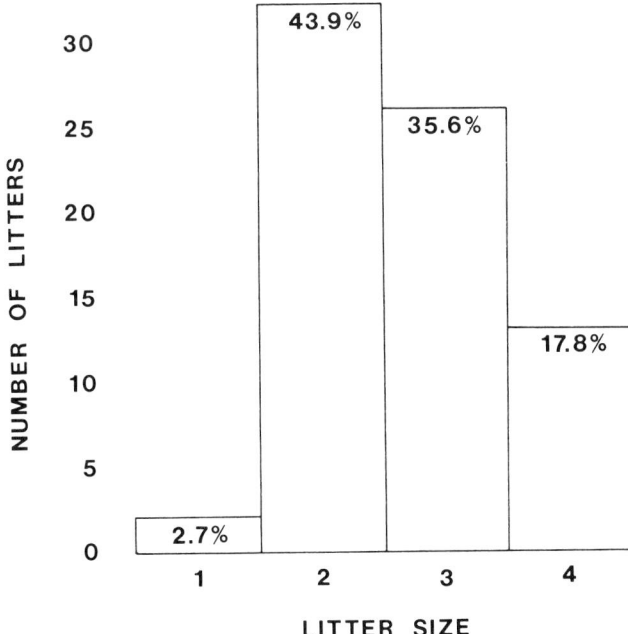

Fɪɢ 11.—Frequency of litter size, with percentage of total for each size based on 38 litters born in captivity and the embryos or fetuses of 35 potential litters.

September, resulting in the recovery of 92 embryos and fetuses, 46 being taken from each of the uterine horns. Embryo counts ranged from one to four, with a mean of 2.63±0.77. The modal litter size based on offspring and embryo counts, was two young (Fig. 11). Placental scars were counted in the reproductive tracts of 63 animals resulting in a mean litter size of 2.81±0.82. Litters, based on scars, ranged from one to four, with a mode of three. Scars were nearly equally divided between the uterine horns, with 83 in the right and 94 in the left.

A summary of litter sizes for *N. micropus* over portions of its geographical range is in Table 4. Litter sizes based on counts of embryos might be slightly high inasmuch as resorption losses are not accounted for had the gestation period been carried to completion. Conversely, the number of young captured with the mother might be low as a result of postnatal mortality. Those records indicate litter size ranges from one to five with a mean generally between 2.2 and 3.0. Rainey (1956) has summarized the data concerning litter size for other species of *Neotoma*.

Determinations of litter size from the present study compare favorably with those listed in Table 4. When litter sizes, as determined from laboratory-born young, embryos from necropsied females, and placental scar counts are combined, the modal litter size is three. Of the 136 litters represented, five contained one young; 51, two young; 54, three young; and 26, four young. Litters in *Neotoma* composed of a single young are few and

TABLE 4.—*Summary of litter size for* Neotoma micropus *as reported in the literature.*

Authority	Locality	Comments	Mean (range)
Feldman (1935)	Eddy Co., New Mexico	Laboratory breeding. Of 11 litters born, 22 young were produced. All litters with two young of same sex.	2.00
Mearns (1907)	Ft. Clark and Ft. Hancock, Texas	Of two females captured, each had three embryos.	3.00
Jonhson (1952)	Zavala Co., Texas	Forty-two pregnant females captured. Number of embryos not given but mean was calculated.	2.25 (?-5)
Raun (1966)	San Patricio Co., Texas	Four females trapped with young. Three with two young and one with a single young. In latter instance, a half-eaten carcass of second young found outside of house.	2.00
Warren (1926)	Baca Co., Colorado	Of two females captured, each had three embryos	3.00
Finley (1958)	Colorado and Oklahoma	Five pregnant females captured with a total of 12 embryos.	2.40 (2-3)
Hall (1955)	Kansas	One to three young produced per litter.	(1-3)
Birney (1970)		Laboratory breeding. Of nine litters born 24 young were produced.	2.66
Birney (1970)	Kansas and Colorado	Sixteen females captured with 41 young.	2.31 (1-4)
	Kansas and Colorado	Eleven females captured which later gave birth to 36 young in laboratory.	3.27 (2-4)
Spencer (1968)		Laboratory breeding. Of 37 litters born, 103 young were produced. Two litters each from crosses between *N. micropus* and *N. floridana*; the remainder from hybrid crosses.	2.78 (1-5)
	Major, Woods, and Woodward Co., Oklahoma	Females captured pregnant, with young, or both. Number of embryos ranged from two to four with a mean of three. Young ranged from one to four with a mean of two.	3.00 (2-4) 2.00 (1-4)
	Major Co., Oklahoma	One female collected, with three near-term embryos.	
	Harmon Co., Oklahoma	Four females captured with from two to four young. One of these were pregnant with four small embryos.	(2-4)
		One captive female had three litters of three, two, and three young.	2.66 (2-3)
	Hamilton Co., Kansas	Three females collected with litters of four, two, and four young.	3.33 (2-4)
Martin and Preston (1970)	Harmon Co., Oklahoma	Adult females captured with one or more young, usually two, rarely one or three.	(1-3)
Baker (1956)	Coahuila	One female captured with two embryos.	
Alvarez (1963)	Tamaulipas	Of two females captured at separate localities, each contained two embryos.	2.00
Dice (1937)	Tamaulipas	Of two females captured, each contained two embryos.	2.00

those containing five occur only rarely. In 100 litters of *N. lepida*, Egoscue (1957) observed only two litters of five. However, *N. cinerea* appears to be an exception. Of 34 litters examined by Egoscue (1962), 12 per cent contained five young and nine per cent, six young.

Potential litter size was determined by counting the number of corpora lutea in the ovaries of 35 pregnant females (see above). A total of 108 corpora lutea was counted for a mean potential litter size of 3.08±0.74. Potential litter size based on such counts ranged from one to five (only one animal possessed five). A 10.2-per cent preimplantation and early postimplantation ovum loss is suggested because of the differential between counts of the corpora lutea (108) and total number of embryos (97).

Of the 35 reproductive tracts examined from pregnant females, five contained one or more resorbing embryos. A reproductive tract of a nonpregnant female was found to contain a single resorbing embryo. It is interesting to note that of the six cases of resorbing embryos observed, four were in reproductive tracts of animals taken in September, the last month of active breeding. Of 97 embryos from pregnant females, five were being resorbed, representing a postimplanatation loss of 5.1 per cent. A combination of pre and postimplantation embryonic losses indicated a postovulatory failure rate of approximately 15 per cent. Since reproductive tracts were examined at all stages of pregnancy, the percentage could have been higher because additional embryos could have resorbed after the time of examination. Based on 19 counts of embryos (mean litter size=3.43), Brown (1969) noted embryo resorption only twice in *N. mexicana*. In each case a single embryo was resorbing high in the uterine horn.

Apparent transmigration of ova has been reported for several species of small mammals by Beer *et al.* (1957) and for *Peromyscus* by Brown (1964, 1966). That was determined by comparing the number of corpora lutea in the right and left ovaries to the number of embryos within the uterine horns. Of 35 pregnant females examined, transmigration of a single ovum was detected in seven, or 20 per cent of the potential litters. That estimate is probably a minimum number, inasmuch as transmigrated ova could be lost from, or exchanged between, horns and not counted.

There were two indications of either monozygotic twinning or polyovuly. In one individual, two corpora lutea were counted in the left ovary, whereas the left uterine horn contained three viable embryos. The right ovary and uterine horn contained no corpora lutea or embryos. In another instance, each ovary contained one corpus luteum, whereas the right uterine horn contained one resorbing embryo and the left, two viable embryos.

Sex Ratios

In the laboratory, 104 young were born to 38 females impregnated in the wild: 47 were males and 57 females. The chi-square test for the deviation from an expected 1:1 ratio gave a nonsignificant ($P>0.05$) value of 0.9614

TABLE 5.—*Sex ratios and chi-square tests for all woodrats captured during the period of study and for those born in the laboratory.*

	Males	Females	Total	χ^2
Young born in laboratory (38 litters)	47	57	104	0.961
Subadults (under 220 grams)	37	84	121	18.256*
Adults (over 220 grams)	192	369	561	55.844*
All woodrats captured (adults and subadults)	229	453	682	73.570*

*Significant ($P<0.005$).

(Table 5). Of more than 200 *N. lepida* born in the laboratory, Egoscue (1957) found an approximate 1:1 sex ratio. Seven *N. floridana* litters born in the laboratory contained 10 males and 11 females (Pearson, 1952).

Throughout my study, 682 woodrats were captured: 229 males and 453 females. As tested by chi-square, the deviation from the expected 1:1 ratio was a highly significant value ($P<0.005$) of 73.570. When the total number of woodrats collected was placed into age classes (adults and subadults), both classes deviated significantly from the expected 1:1 ratio. Chi-square values for those classes were 55.884 and 18.256, respectively, and highly significant ($P<0.005$). In the population as a whole, there existed a 1:1.98 ratio of males to females.

An analysis of Table 5 demonstrates that with increasing age the deviation from a 1:1 ratio increases in favor of more females, the implication being that with increasing age males are more subject to higher mortality losses than are females. Although the possiblity of differential trapping success cannot be entirely discounted for the noted difference, it seems unlikely that this is the determining factor. A more plausible explanation for an unbalanced sex ratio is that males are subject to higher predation pressure as a result of their generally larger homes ranges and longer movements in search of estrous females. Cranford (1977) noted an increase in the home range of male *N. fuscipes* during those months in which females were in estrus. Such movements, which undoubtedly involve some marginal habitat, increase their exposure to predators. That males make longer movements than do females has been well documented by a number of investigators (for example, Raun, 1966, for *N. micropus*; and Fitch and Rainey, 1956, for *N. floridana*). Nearly equal sex ratios have been reported from large samples of other species of woodrats (see Johnson, 1952, for *N. micropus*; Goertz, 1970, and Rainey, 1956, for *N. floridana*; Linsdale and Tevis, 1951, for *N. fuscipes*; and Vorhies and Taylor, 1940, for *N. albigula*).

Number of Litters per Year

Most determinations of the number of litters produced per year by a female are based on live-trapping, mark-release studies. Females can be examined periodically throughout the year in this manner. The number of litters born per year in the laboratory is also utilized. In my investigation, the woodrat population on the study area was sampled at monthly intervals, and data on the determination of the number of litters per year were acquired by indirect means.

Of 35 pregnant woodrats necropsied, four were in lactating condition and nine in postlactating condition, indicating at least two litters are produced per year. No doubt, a number of the other pregnant females were in at least their second pregnancy, but, if a sufficient time lag existed betwen the previous litter and the current pregnancy, the nipples could have returned to the nonlactating condition. When such cases exist, it is not possible to determine what litter the current pregnancy represents.

The production of at least two litters per year also was indicated by the number of pregnant females live-trapped and brought to the laboratory in a lactating or postlactating condition. Of 38 such pregnant females, two were lactating and six were postlactating when captured.

An examination of placental scars lends additional evidence to the number of litters produced by individual females. Most placental scars cannot be seen in pregnant animals as the developing embryos obscure part or all of them, especially during latter stages of gestation. When observed, they are usually in the horn containing no embryos. Eleven pregnant females captured from June through September contained placental scars. Therefore, these females were producing at a minimum their second litter. One pregnant female contained four distinct scars and a second set of older ones, thus indicating the production of at least three litters.

A conservative estimate of yearly litter production can be gained from an examination of scars in reproductive tracts of nonpregnant females. In 17 tracts, two contrasting sets of scars, old and new, were observed. This indicated at least two litters had been produced by the females prior to capture. One animal captured in July had three new, and seven old, scars. Even if some of the most recent scars were superimposed upon older ones, that female possessed 10 scars representing at least three litters, inasmuch as litter size rarely exceeds four young.

Other reproductive tracts contained numerous distinct scars that could not be differentiated as either old or new. Those tracts contained more scars than would be produced by a single litter. Seventeen such tracts were observed from August through March; most contained from five to seven scars. Those scars represented young born during the last breeding season. Tracts containing 10 (October), 13 (January), nine and 15 (February) placental scars were observed. The tract containing 15 distinct scars most likely represented four litters produced the previous season, dependent upon litter size.

At least two litters per year are produced based on placental scar analysis. However, due to the life span of placental scars, it is somewhat more difficult to assign the production of more than two litters per year to an animal. Because some females were producing their second litter as early as June, this suggests ample time for additional litters. My data suggest many females produced a minimum of two litters per year and that at least some (possibly many) produced more.

My data are in agreement with the findings of other investigators. Raun (1966) indicated that *N. micropus* gave birth to two to three litters per year and that some produced five. According to Spencer (1968), females probably have up to three litters per year. Based on dates of capture of young woodrats and breeding females, Finley (1958) thought two or more litters were produced yearly under favorable conditions. Warren (1926) indicated at least two litters were produced per year.

Gestation Periods

Limited observations were obtained on the gestation period of *N. micropus*. In the laboratory, breeding trials were accomplished by introducing a female into a breeding cage containing a male. If no fighting occurred, the pair was left together for a week. As a result of such pairings, three litters were born. For one female, the minimal and maximal limits were 30 to 37 days, and, for another female, 31 to 38 days. A third mating, in which the pair was left together for two weeks, resulted in the birth of a litter 39 days after the pair was separated. The gestation period might have been longer. A female trapped on 30 July 1971 gave birth to a litter of four young 33 days later, establishing a minimal gestation period of 33 days in that instance.

The gestation period of *N. micropus* appears somewhat variable. Feldman (1935) determined the gestation period for the species to be less than 33 days. Spencer (1968) visually recorded the time of copulation, sighting of litters within 24 hours of parturition, and calculated the gestation period of two litters at 33 and 38 days. Neither Birney (1970) nor Spencer (1968) observed any sharp differences in gestation periods between *N. floridana* and *N. micropus*. Birney (1970) reported gestation period probably fluctuating between 32 to 38 days with a modal duration of 35 days. Of 35 intra and interspecific crosses (four intraspecific) made by Spencer (1968), the mean gestation period was calculated at 35.1 days.

Spencer (1968) crossed females of *N. floridana* and *N. micropus* in post-partum estrus to F_1 hybrids between those species. Those females gave birth to litters 51 and 55 days later. Spencer captured a female *N. micropus* with young three to nine days old. She bore a litter 46 days later although never paired with a male while in captivity. He calculated the second gestation period from 49-54 days because of copulation during a postpartum estrus. Feldman (1935) reported the presence of a postpartum estrus in *N. albigula*. Pregnant, lactating females were caught on several occasions, but no positive evidence was gained in this study either to substantiate or to refute the presence of a postpartum estrus.

Parturition

No observation was made of a female giving birth to young. A caged female was discovered that had only recently given birth to two young at 930 hours on 31 May 1971. One young was attached to a nipple whereas the other was lying in the wood shavings nearby. The placenta of the latter was still attached by the umbilical cord. The young, placenta, and fetal membranes were still wet. The female was somewhat docile and reluctant to move. The cage was checked again at 1000 hours, and both young were attached to nipples and sucking. The umbilical cord had been bitten off close to the body and eaten along with the placenta. At 1230 hours the cage again was checked, and a third young had been born. It was attached to a nipple and still possessed 2.5 centimeters of the umbilical cord although the placenta was missing. One ear of the first-born young was unfolding at this time.

No placentae were ever found with the exception of this observation. Evidently the placentae, fetal membranes, and umbilical cords are eaten almost immediately after parturition. The placentae of *N. micropus* (also observed on removal of near-term embryos) are similar to those of *N. floridana* described by Knoch (1968).

Although no direct observations of parturition have been recorded in the literature for *N. micropus*, details likely resemble those recorded for *N. floridana* (Rainey, 1956; Murphy, 1952), *N. lepida* (Rainey, 1965), and *N. albigula* (Bleich and Schwartz, 1975).

Vaginal Plugs

There are a few reports in the literature on observation of vaginal plugs (= copulation plugs). Rainey (1956) reported the capture of a female *N. floridana* in the same trap with a male in breeding condition. The female possessed a vaginal plug at this time and upon recapture 33 days later, embryos were detected by palpation. Rainey also mentioned the presence of a vaginal plug for one other female.

Vaginal plugs, a coagulum of fluid from the male's vesicular and coagulating glands, form in the vagina as a result of recent copulation (Rugh, 1968). However, Spencer (1968) examined females of *N. micropus*, *N. floridana*, and their hybrids immediately following copulation and at random intervals up to 24 hours after copulation and found no vaginal plugs at that time or during his other studies of the two species. No vaginal plugs were noticed by Wood (1935) in breeding experiments with *N. fuscipes*.

In my study, vaginal plugs were detected only during the months of April to August. Plugs were a white-colored, translucent, granular semisolid, which completely filled the vagina. The distal end was expanded somewhat. On several occasions, a plug was grasped with forceps and gently pulled outward 6 to 9 millimeters. When released it quickly retracted back into the vagina. Raun (1966) made similar observations and found vaginal plugs in nine females. He recaptured one female three days after observance of a plug, but it had disappeared; another female estimated to be four months old, and apparently infertile, had a vaginal plug.

Of the females necropsied, only two possessed vaginal plugs. In both instances, the plug material completely filled the vagina, body of the uterus, and uterine horns. The uterine horns were turgid and almost transparent. Neither reproductive tract contained visible embryos. By examination of the nipples, one animal was lactating, whereas the other was not. The milk glands of the former were highly developed and contained much milk. Evidently, this animal had copulated either in postpartum estrus or during lactation.

Only one of five females with vaginal plugs at time of capture (three in lactating, one in postlactating, and one in nonlactating condition) later produced a litter. That female, in lactating condition, gave birth to two young 19 days after the vaginal plug was first noticed. If the vaginal plug resulted from copulation, then it must have persisted since conception. The most recent pregnancy must have resulted from copulation during a postpartum estrus or during lactation. However, the litter was not born late enough after capture to confirm a gestational extension resulting from lactation.

The vaginae of four females were found to contain a very thick reddish material of low viscosity. However, unlike vaginal plugs, this material was not a semisolid. Three of those animals were killed and found to have small embryos; the largest was 19 millimeters in crown-rump length. One individual had just completed lactation, as there was a trace of milk remaining in the milk glands. The fourth was returned to the laboratory and gave birth to three young 19 days later. The function, or significance, of the reddish material is unknown.

Ventral Abdominal Gland

Size of the ventral abdominal gland and the stained coloration of the pelage along its lateral sides were studied and recorded. Information was obtained from 280 *N. micropus* collected at 12 monthly intervals in an effort to correlate the development of the ventral abdominal gland with the breeding season.

The ventral abdominal gland of male *N. micropus* occupies a region of sparsely haired skin that extends along the ventral midline from the posterior edge of the ribcage to the umbilicus. In rare instances, the gland extends up to 12 millimeters posterior to the umbilicus. In adult males, it measures 3 to 9 millimeters in width and 40 to 65 millimeters in length. The surface texture appears rough, scaley, and glandular, especially in large, old males. Hair adjacent to the gland normally overlaps and covers it, but does not when the gland is large and the surface highly developed. In females, the gland is much smaller and measures 2 to 5 millimeters in width and 20 to 55 millimeters in length. The gland is less well developed than it is in males, and the surface texture is rough, scaly, and nonglandular. Hair grows from the surface of the gland in females. Often the gland is faint and barely visible and in some instances lacking. However, one female taken on 29 June 1971 possessed a ventral abdominal gland measuring 8 by 61 millimeters and was as well developed as that of most males. No staining of the adjacent pelage was noticeable.

In males, the exudate from the gland faintly stains the pelage along the lateral edges of the gland in a band six to nine millimeters wide. At no time throughout the year could a detectable difference be observed on the development of the gland or its staining effect. No staining effect of the pelage was ever noticed in females. Spencer (1968), Egoscue (1962), and Linsdale and Tevis (1951), studying N. micropus and N. floridana, N. cinerea, and N. fuscipes, respectively, noted that the ventral abdominal gland was most active during the breeding season. My observations of N. floridana in Kansas (Wiley, 1971) are in agreement with the above investigators; N. floridana has a very active ventral abdominal gland during the breeding season, so much so, in fact, that occasionally almost the entire venter becomes darkly stained. Egoscue (1962) made similar observations of the intense staining effect on the pelage of male N. cinerea.

Spencer (1968) and Egoscue (1972) noted that males introduced into a new cage would deliberately drag their venter over objects within the cage. Spencer attributes this behavior to a "sign posting" activity. This is probably involved with territory marking and sex recognition. Howe (1977) indicated that scent marking by woodrats has several social functions involving sexual and agonistic behavior. During the present study, venter dragging by males introduced into unoccupied breeding cages was not observed. From year-around observations of the ventral abdominal gland, no correlation was obvious between its development, staining effect, and the season of breeding. The most indepth study of the ventral abdominal gland was conducted by Clarke (1973) for N. floridana.

Summary

The reproductive biology of the southern plains woodrat, Neotoma micropus, was studied from October 1970 through September 1971. Woodrats studied were from near Wink, Winkler County, Texas.

The season of breeding, as determined by the presence of embryos, birth of young, and examination of mammae extended from March to mid-

October. Breeding was most intense from June to September, when 42 per cent of all adult females sampled were pregnant. Initiation and termination of the breeding season was synchronous. Nipple condition was a reliable indicator of the breeding season, whereas examination of the vaginal orifice was of limited use. A minimum of 90 per cent of the adult females bred during the year.

Testes varied little in weight throughout the year, but weighed less from October to January than at other times. As a reflection of rather constant testes weights, epididymides contained abundant motile sperm year-around. Most scrotal males occurred during months when females were breeding. Seminal vesicle weights were greatest in July and were least in weight from October to March. Greatest seminal vesicle weights corresponded closely to the season of breeding. About 30 per cent of the ovaries examined during the months of nonbreeding contained active corpora lutea of ovulation. The presence of testicular and epididymal sperm and some ovaries with corpora lutea of ovulation during the months of nonbreeding suggest ethological barriers to breeding. The decrease in seminal vesicle weights, and hence testosterone levels during the winter months, might suppress the male sex drive (thus preventing breeding).

Mean litter size was 2.73±0.83 based on laboratory born litters, 2.63±0.77 based on embryo counts, and 2.81±0.82 based on placental scar counts. Litter size ranged from one to four, with a mode of three. Mean potential litter size based on corpora lutea counts was 3.08±0.74. Pre and postimplantation losses accounted for over 15 per cent of all ova ovulated. Transmigration of ova was detected in seven of 35 pregnant females.

Sex ratios at birth approximated a 1:1 ratio. Among adults, there were significantly more females than males, suggesting that males were subject to greater predation pressure as a result of their larger home ranges.

Adult females produced a minimum of two litters per year, with indications that some might produce as many as four. Based on year-around observations of the ventral abdominal gland, no correlation was obvious between its development, histological appearance, or the season of breeding.

Several conclusions can be postulated based on these findings. *Neotoma micropus* has evolved a seasonal reproductive cycle, as have other temperate-zone members of the genus, and cricetine rodents in general. Such a cycle, even in such southerly latitutdes as Texas, results in litter production during those times of the year conducive to litter survival. The possibility exists, however, that breeding might be noncyclic during years of little environmental (temperature-moisture) fluctuation. Physical environmental factors either singly or in interaction, might act directly on the population in establishing the breeding season. However, there were indications that the indirect action of these factors, acting through plants, nutritionally might control commencement, duration, and termination of the breeding season. The beginning and end of the annual reproductive cycle appeared to be correlated closely with the major phenological changes of the vegetation at its initiation and termination. Additional studies of the effect of nutrition on the breeding season of *N. micropus* would likely prove productive.

Most males and approximately a third of the females examined possessed breeding capabilities throughout the nonbreeding portion of the year. Ethological factors probably prevented those individuals from reproducing at that time. That the reproductive organs of a portion of the population are functional year-around might be of two-fold significance. First, it could permit an accelerated rate of recruitment in times of low population levels. That would be adaptively significant to the population in allowing rapid recovery from low densities produced by environmental resistance. Secondly, because the population remains reproductively ready at all times, it could avoid energy dissipation in activation of reproductive tissue. Energy thus saved from the reproductive processes could be utilized by the population in other body functions and activities.

The litter size and number of litters produced per year by *N. micropus* obviously balances natality with environmental losses to maintain the population in a dynamic equilibrium. Size of litters and numbers per year determined in this study compare favorably with those of the species from other geographical localities. Further study of litter size and number of litters per year might reveal slight latitudinal differences intraspecifically as noted in some cases interspecifically with latitude. If such differences exist, they naturally would tend to adapt the populations' litter size and timing of litters to the local environmental conditions. All of these factors would serve to enhance the adaptability of the species, and could account for its success in such a large geographic area.

ACKNOWLEDGMENTS

I am especially indebted to Dr. Robert L. Packard, who directed this research and aided in preparation of the manuscript. Drs. Eric Bolen, Francis Rose, Robert Baker, and John Mecham read the manuscript and made valuable suggestions. Dr. Francis Rose was helpful in providing laboratory space, and Dr. Robert Baker in the provision of field and laboratory equipment. Useful help was provided by Drs. Larry Brown, John Burns, and Willie Bleier in examination and interpretation of histological preparations. Drs. Philip Morey, Rick McDaniel, and Tony Mollhagen aided in photography, and Dr. Dede Armentrout assisted with wax embedding and sectioning techniques. I am grateful to Drs. Robert Martin and Kenneth Matocha, and my wife Suzanne, for helpful suggestions throughout the investigation and for the care of laboratory animals during my absence.

LITERATURE CITED

ALVAREZ, T. 1963. The recent mammals of Tamaulipas, Mexico. Univ. Kansas Publ., Mus. Nat. Hist., 14:363-473.

ANDERSON, S. 1969. Taxonomic status of the woodrat, *Neotoma albigula* in southern Chihuahua, Mexico. Pp. 25-50, in Contributions in mammalogy—a volume honoring Professor E. Raymond Hall (J.K. Jones, Jr., ed.), Univ. Kansas Mus. Nat. Hist., Misc. Publ., 51:1-428.

BAILEY, V. 1931. Mammals of New Mexico. N. Amer. Fauna, 53:1-399.

BAKER, R. H. 1956. Mammals of Coahuila, Mexico. Univ. Kansas Publ., Mus. Nat. Hist., 9:125-335.

BEER, J. R., C. F. MacLEOD, AND L. D. FRENZEL. 1957. Prenatal survival and loss in some cricetid rodents. J. Mamm., 38:392-402.

BIRNEY, E. C. 1970. Systematics of three species of woodrats (genus *Neotoma*) in central North America. Unpublished Ph.D. dissertation, Univ. Kansas, Lawrence.

BLAIR, W. F. 1943. Populations of the deer-mouse and associated small mammals in the mesquite association of southern New Mexico. Contrib. Lab. Vert. Biol., Univ. Mich., 21:1-40.

——. 1950. The biotic provinces of Texas. Texas J. Sci., 2:93-117.

BLEICH, V. C., AND O. A. SCHWARTZ. 1975. Parturition in the white-throated woodrat. Southwestern Nat., 20:271-272.

BROWN, L. N. 1964. Reproduction of the brush mouse and white-footed mouse in the central United States. Amer. Midland Nat., 72:226-240.

——. 1966. Reproduction of *Peromyscus maniculatus* in the Laramie Basin, Wyoming. Amer. Midland Nat., 76:183-189.

——. 1969. Reproductive characteristics of the Mexican woodrat at the northen limit of its range in Colorado. J. Mamm., 50:536-541.

CLARKE, J. W. 1973. The specialized midventral gland of the eastern woodrat, *Neotoma floridana osagensis*. Unpublished M.S. thesis, Kansas State Teachers College, Emporia, 68 pp.

CRANFORD, J. A. 1977. Home range and habitat utilization by *Neotoma fuscipes* as determined by radiotelemetry. J. Mamm., 58:165-172.

DAVIS, W. B. 1974. The mammals of Texas. Bull. Texas Parks and Wildlife Dept., 41 (revised):1-294.

DICE, L. R. 1937. Mammals of the San Carlos Mountains and vicinity. Univ. Mich. Studies, Sci. Ser., 12:245-268.

DONAT, F. 1933. Notes of the life history and behavior of *Neotoma fuscipes*. J. Mamm., 14:19-26.

EGOSCUE, H. J. 1957. The desert woodrat: a laboratory colony. J. Mamm., 38:472-481.

——. 1962. The bushy-tailed wood rat: a laboratory colony. J. Mamm., 43:328-377.

ENGLISH, P. F. 1923. The dusky-footed wood rat (*Neotoma fuscipes*). J. Mamm., 4:1-9.

FELDMAN, H. W. 1935. Notes on two species of wood rats in captivity. J. Mamm., 16:300-303.

FINLEY, R. B., JR. 1958. The wood rats of Colorado: distribution and ecology. Univ. Kansas Publ., Mus. Nat. Hist., 10:213-552.

FITCH, H. S., AND D. G. RAINEY. 1956. Ecological observations on the woodrat, *Neotoma floridana*. Univ. Kansas Publ., Mus. Nat. Hist., 8:499-533.

GANDER, F. F. 1929. Experiences with wood rats, *Neotoma fuscipes macrotis*. J. Mamm., 10:52-58.

GARNER, H. W. 1974. Population dynamics, reproduction, and activities of the kangaroo rat, *Dipodomys ordii*, in western Texas. Grad. Studies, Texas Tech Univ., 7:1-28.

GOERTZ, J. W. 1970. An ecological study of *Neotoma floridana* in Oklahoma. J. Mamm., 51:94-104.

GOLLEY, F. B. 1962. Mammals of Georgia. Univ. Georgia Press, Athens, 218 pp.

HALL, E. R. 1955. Handbook of mammals of Kansas. Univ. Kansas Mus. Nat. Hist., Misc. Publ., 7:1-103.

————. 1981. The mammals of North America. 2nd ed. John Wiley & Sons, New York, 2:viii+601-1181+90.

HAMILTON. W. J., JR. 1953. Reproduction and young of the Florida wood rat, *Neotoma f. floridana* (Ord). J. Mamm., 34:180-189.

HERSH, S. L. 1981. Ecology of the Key Largo woodrat (*Neotoma floridana smalli*). J. Mamm., 62:201-206.

HOLDENRIED, R., AND H. B. MORLAN. 1956. A field study of wild mammals and fleas of Santa Fe County, New Mexico. Amer. Midland Nat., 55:369-381.

HOWE, R. J. 1977. Scent-marking behavior in three species of woodrats (*Neotoma*) in captivity. J. Mamm., 58:685-688.

JOHNSON, C. W. 1952. The ecological life history of the packrat, *Neotoma micropus*, in the brushlands of southwest Texas. Unpublished M.A. thesis, Univ. Texas, Austin, 115 pp.

KNOCH, H. W. 1968. The eastern wood rat, *Neotoma floridana osagensis*: a laboratory colony. Trans. Kansas Acad. Sci., 71:361-372.

LINSDALE, J. M., AND L. P. TEVIS, JR. 1951. The dusky-footed wood rat. Univ. California Press, Berkeley, x+664 pp.

MARTIN, R. E., AND J. R. PRESTON. 1970. The mammals of Harmon County, Oklahoma. Proc. Oklahoma Acad. Sci., 49:42-60.

MEARNS, E. A. 1907. Mammals of the Mexican boundary of the United States. Bull. U. S. Nat. Hist. Mus., 56(1):1-530.

MURPHY, M. F. 1952. Ecology and helminths of the Osage wood rat, *Neotoma floridana osagensis*, including the description of *Longistriata neotoma* n. sp. (Trichostrongylidae). Amer. Midland Natur., 48:204-218.

NALBANDOV, A. V. 1976. Reproductive physiology of mammals and birds. W. H. Freeman and Co., San Francisco, xv+344 pp.

PEARSON, P. G. 1952. Observations concerning the life history and ecology of the woodrat *Neotoma floridana floridana* (Ord). J. Mamm., 33:459-463.

POOLE, E. L. 1940. A life history sketch of the Allegheny woodrat. J. Mamm., 21-249-270.

RAINEY, D. G. 1956. Eastern woodrat, *Neotoma floridana*: life history and ecology. Univ. Kansas Publ., Mus. Nat. Hist., 8:535-646.

————. 1965. Parturition in the desert wood rat. J. Mamm., 46:340-341.

RAUN, G. G. 1966. A population of woodrats (*Neotoma micropus*) in southern Texas. Bull. Texas Memorial Mus., 11:1-62.

RICHARDSON, W. B. 1943. Wood rats (*Neotoma albigula*): their growth and development. J. Mamm., 24:130-143.

RUGH, R. 1968. The mouse, its reproduction and development. Burgess Publishing Co., Minneapolis, iv+430 pp.

SHORT, R. V. 1972. Role of hormones in sex cycles. Pp. 42-72, *in* Hormones in reproduction (C. R. Austin and R. V. Short, eds.). Cambridge Univ. Press, Cambridge, viii+148 pp.

SPENCER, D. L. 1968. Sympatry and hybridization of the eastern and southern plains wood rats. Unpublished Ph.D. dissertation, Univ. Oklahoma, Stillwater, 85 pp.

TINKLE, D. W., D. McGREGOR, AND S. DANA. 1962. Home range ecology of *Uta stansburiana stejnegeri*. Ecology, 43:223-229.

VESTAL, E. H. 1938. Biotic relations of the wood rat (*Neotoma fuscipes*) in the Berkeley Hills. J. Mamm., 19:1-36.

VORHIES, C. T., AND W. P. TAYLOR. 1940. Life history and ecology of the white-throated wood rat, *Neotoma albigula* Hartley in relation to grazing in Arizona. Univ. Arizona, Coll. Agric., Tech Bull., 86:455-529.

WARREN, E. R. 1926. Notes on the breeding of wood rats of the genus *Neotoma*. J. Mamm., 7:97-101.

WILEY, R. W. 1971. Activity periods and movements of the eastern woodrat. Southwestern Nat., 16:43-54.

———. 1972. Reproduction, postnatal development, and growth of the southern plains woodrat, (*Neotoma micropus*) in western Texas. Unpublished Ph.D. dissertation, Texas Tech Univ., Lubbock, 142 pp.

WOOD, F. D. 1935. Notes on the breeding behavior and fertility of *Neotoma fuscipes macrotis* in captivity. J. Mamm., 16:105-109.

BEHAVIOR OF NORTH AMERICAN GEOMYIDS DURING SURFACE MOVEMENT AND CONSTRUCTION OF EARTH MOUNDS

GRAHAM C. HICKMAN

A number of unrelated and allopatric fossorial mammals, such as the African Bathyergidae (Hickman, 1979a) and South American Ctenomyidae (Pearson, 1959), produce scattered surface mounds of loose earth. In North America, pocket gophers (Geomyidae) also construct earth mounds (Hickman, 1977a), which influence both soil and vegetation (Mielke, 1977).

The literature on the mound-building of geomyids is largely anecdotal. Mound-building behavior of the southeastern pocket gopher *Geomys pinetis*, as studied in some detail by Hickman and Brown (1973a), concurs with casual descriptions of the mound-building behavior of several species of *Thomomys* (Wight, 1942; Seton, 1929) and the plains pocket gopher, *Geomys bursarius* (Breckenridge, 1929; Scheffer, 1910), although there are some distinct differences. The major features of mound-building for *Pappogeomys castanops* have been mentioned briefly (Hermann, 1950), but no major study of mound construction in that species has been conducted.

The factors involved in the location, amount, and manner of construction of earth mounds remains largely unscrutinized for most fossorial mammals. The present study involves a comparative analysis of mound-building behavior of three species (*Geomys bursarius*, *Thomomys bottae*, and *Pappogeomys castanops*) representing each of the three genera of North American geomyids under varying conditions of ground cover, soil types, moisture, temperature, and vulnerability to disruption by other organisms. Once the major behavior patterns of mound-building have been identified qualitatively and quantitatively, the adaptive significance of inter and intraspecific differences may become apparent. Moreover, a comparison of the less flexible behavioral patterns may contribute to an understanding of the phyletic affinities of the three genera to the tribes Thomomyini and Geomyini, as outlined by Russell (1968).

MATERIALS AND METHODS

Pocket gophers were captured with Baker and Williams (1972) live-traps: eight *Geomys bursarius* from north of Slaton, Lubbock County, Texas; six *Pappogeomys castanops* from Lubbock, Texas; and eight *Thomomys bottae* from the vicinity of Sacramento, Otero County, New Mexico. Animals were maintained in captivity in 20 gallon garbage cans half-filled with sandy soil and fed crimped oats and laboratory rabbit chow with occasional supplements of lettuce.

165

Individual pocket gophers were tested between 0800 and 1700 hours once a day in areas in the vicinity of Lubbock, Texas, from March through June, and in September of 1973. Areas were selected so as to provide variation in conditions of soil composition, moisture conditions, and amounts of ground cover needed to evoke any varied behavior repertoire involved in mound-building, and thereby determine what factors influence digging ability and style. Although some animals were released in the same areas, conditions could have varied greatly, particularly before and after a rain.

Mound building ethology has been divided into five phases (Hickman and Brown, 1973a):

Prospecting involves exploratory behavior involved in the selection of a digging site, beginning with the tipping over of a transporting cage and ending with the animal digging and first pushing soil out of the excavation. The observer maintained a distance of three meters from the animal while mapping length and direction of paths taken and time expired, so as not to influence the proclivity of the animal to dig.

Ground-breaking encompasses the manner in which a mound is initiated once a digging site is chosen. At this point, the observer was able to approach slowly to within a meter of the pocket gopher without any apparent disruption to digging. Initial compass directions in which tunnels were initially excavated were noted.

Excavation of the tunnel by loosening of soil with teeth and claws is the next phase. The direction of tunnels excavated following initial penetration of the surface, and direction of turns (left or right) was recorded. A "turn" refers to the action of turning away from the excavation site and facing the burrow entrance to thrust tailings out to the surface. The total number of turns could not always be tallied for each excavation inasmuch as pocket gophers in the field often attained depths limiting visual monitoring.

Distribution (=Mound-building proper) concerns the pushing of soil to the surface and distribution of the tailings about the burrow entrance. The direction (left, straight, or right) and number of thrusts were recorded. This phase alternates with the excavation phase and ends with a "return" (the movement patterns of turning away from the tunnel opening to face the end of the tunnel preliminary to further excavation).

Plugging of the burrow entrance with soil completes mound-buiding unless the tunnel is reopened and more dirt added to form a "multiple mound." The total time from ground-breaking to plugging was noted.

Both "turns" of the excavation phase and "returns" of the mound-building phase have been termed "about faces," the stereotypic method of making a 180° turn in the burrow. It is important to note that tunnel direction and turns have been indicated as being either to the "left" or "right" of the pocket gopher while the animal was excavating; alternatively, dirt-thrusts and returns were recorded as being either to the "left" or "right" as the pocket gopher faced the entrance of the burrow.

Field observations of the excavation, distribution, and plugging phases were supplemented by observing pocket gophers in an observation chamber (5.4 meters long, 1.2 meters high, and 7.2 centimeters wide), where observations of subsurface activity could be made in tunnels next to the glass (see Hickman, 1978, for a review of various techniques for observing fossorial mammals in captivity).

RESULTS

Prospecting

Upon release on the surface, *Geomys bursarius*, *Pappogeomys castanops*, and *Thomomys bottae* wandered randomly, recrossing their own pathways several times. The "stalking" movements during prospecting were subtle, with very little lateral or vertical movement of the body, as when locomoting through tunnels (Fig. 1). As a result, pocket gophers were difficult to detect traversing through even low grassy areas.

Few of the habitat conditions at release sites appeared to act as barriers during prospecting. One *Geomys bursarius* released after a rain waded through the edge of a puddle for approximately two meters, avoiding any depth greater than several centimeters, whereas another paused to drink from a puddle. Pocket gophers readily crossed eight meters or more of pavement in the course of exploration. One *Geomys bursarius* wandered near a ground squirrel (*Spermophilus tridecemlineatus*) burrow and was confronted by the occupant. The pocket gopher lunged and landed on the ground squirrel causing it to squeal and retreat to the safety of its burrow. Inquisitive sparrows elicited no visible response even when approaching within 1.3 meters of each species.

Quantitative data regarding prospecting is presented in Table 1. The longest distance moved in prospecting by each species was: *Geomys bursarius*, 204 meters (mean, 37 meters); *Pappogeomys castanops*, 168 meters (mean, 42 meters); *Thomomys bottae*, 156 meters (mean, 23 meters). The longest time spent in prospecting was 41 minutes (mean, 7 minutes) for *Geomys bursarius*, 60 minutes (mean, 15 minutes) for *Pappogeomys castanops*, and 15 minutes (mean, 6 minutes) for *Thomomys bottae*; at least one individual of each species began excavating immediately upon release. The greatest surface speed for *Geomys bursarius* was 31 meters per minute (mean, 4.8), as compared with 9.3 meters per minute (mean, 2.0) for *Pappogeomys castanops*, and 15 meters per minute (mean, 3.4) for *Thomomys bottae*. All three species paused in areas of high, dense vegetation.

Trees and bushes did not appear to attract or detract from digging locations, although some pocket gophers lingered up to five minutes in the shade of trees. Ambient temperature during prospecting was variable enough for comparisons during dry conditions only for *Geomys*, and then only at the 10° to 18°C range. Meters traveled fell 42 per cent during dry soil trials and from a mean of 49.8 to 29.1 (6 trials), minutes 30 per cent (from a

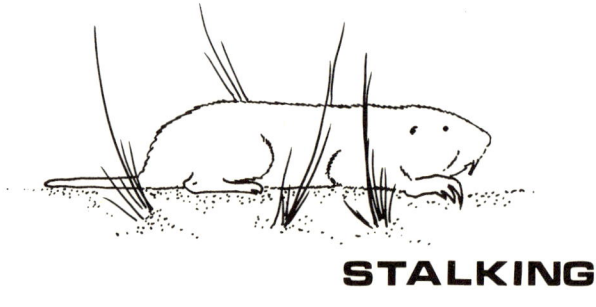

STALKING

CROUCHING

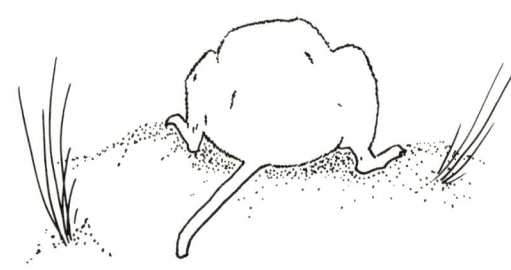

RIGID-TAIL

Fig. 1.—Postures of pocket gophers while prospecting and mound-building as demonstrated by *Geomys bursarius.*

mean of 10.3 to 7.3), and meters per minute 17 per cent (from a mean of 4.8 to 4.0).

Similar values for prospecting rate in meters per minute were obtained at the 10° to 18°C range during moist conditions for *Thomomys* (4.2, 13 trials), and *Pappogeomys* (4.0, 9 trails). All three species decreased the number of meters traveled per minute at the higher 21° to 29°C level during moist conditions according to species size (*Thomomys*, 24; *Geomys*, 35; *Pappogeomys*, 50 per cent); as the number of meters traveled decreased (*Thomomys*, 69; *Geomys*, 98; *Pappogeomys*, 22 per cent), minutes prospecting decreased slightly (11 per cent) for *Thomomys* and *Geomys* but increased by 65 per cent for *Pappogeomys.*

TABLE 1.—Rate of movement on the surface for prospecting Geomys bursarius, Pappogeomys castanops, and Thomomys bottae under varying conditions of temperature and soil moisture. Abbreviations: male (M), female (F), dry (D), moist (M), meters (m), minutes (min.), mean ($\bar{X}$), standard deviation (SD).

Geomys bursarius						Pappogeomys castanops						Thomomys bottae					
Grams/Sex	Dry-moist	°C	m.	Min.	m/Min.	Grams/Sex	Dry-moist	°C	m.	Min.	m/Min.	Grams/Sex	Dry-moist	°C	m.	Min.	m/Min.
75M	M	13	0.5	0.5	0.5	114F	M	24	76.0	60.0	1.2	59F	M	16	156.0	11.0	14.2
75M	M	27	0.5	0.5	0.5	218M	M	24	0.5	0.5	0.5	60F	M	16	0.5	0.5	0.5
75M	M	27	0.5	0.5	0.5	218M	M	16	0.5	0.5	0.5	60F	M	18	5.0	4.0	1.3
75M	M	22	0.5	0.5	0.5	218M	D	16	55.0	39.0	1.4	60F	M	24	0.5	0.5	0.5
105F	M	22	17.0	4.0	4.3	283M	D	16	0.5	0.5	0.5	60F	M	16	3.0	2.0	1.5
173F	M	10	41.0	8.0	5.1	283M	M	16	0.5	0.5	0.5	69F	M	16	43.0	8.0	5.4
173F	M	10	27.0	12.0	2.3	310F	M	24	20.0	26.0	0.8	69F	M	18	12.0	5.0	2.4
173F	D	13	0.5	0.5	0.5	310F	M	18	0.5	0.5	0.5	69F	M	24	27.0	7.0	3.9
180F	D	18	0.5	0.5	0.5	310F	M	16	3.0	5.0	0.6	75F	M	16	0.5	0.5	0.5
183F	M	16	27.0	17.0	1.6	310F	M	29	0.5	0.5	0.5	87M	M	42	6.0	3.0	2.0
183F	M	22	0.5	0.5	0.5	310F	D	27	0.5	0.5	0.5	88M	M	16	3.0	1.0	3.0
223M	D	22	204.0	20.0	10.2	316F	M	16	0.5	0.5	0.5	88M	M	18	5.0	15.0	0.3
223M	D	10	0.5	0.5	0.5	320F	M	18	168.0	18.0	9.3	88M	M	24	9.0	7.0	1.3
223M	M	27	0.5	0.5	0.5	320F	M	24	125.0	34.0	3.7	88M	M	16	8.0	11.0	0.7
228F	D	10	0.5	0.5	0.5	320F	M	18	78.0	34.0	2.3	94F	M	16	31.0	7.0	4.4
269M	M	10	116.0	15.0	7.7	320F	M	29	46.0	9.0	5.1	94F	M	23	49.0	11.0	15.0
269M	D	10	0.5	0.5	0.5	335F	M	16	149.0	20.0	7.5	94F	M	16	1.0	3.0	0.3
269M	M	13	87.0	9.0	31.0	353M	M	24	12.0	10.0	1.2	98M	M	16	78.0	15.0	5.2
269M	D	13	172.0	41.0	4.2	353M	M	18	63.0	38.0	1.6	98M	M	23	6.0	3.0	2.0
$\bar{X} \pm$ SD		37±14.2		7±2.4	4.8±1.7	$\bar{X} \pm$ SD		42±13.0		15±4.2	2.0±0.6	$\bar{X} \pm$ SD		23±8.7		6±1.1	3.4±1.0

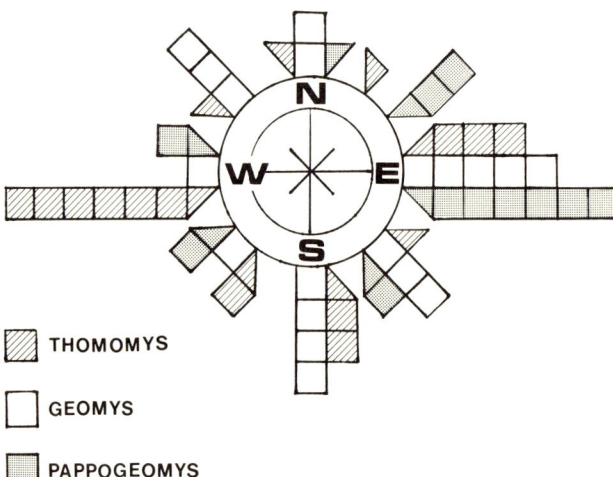

THOMOMYS

GEOMYS

PAPPOGEOMYS

Fig. 2.—Compass directions in which tunnels were initially excavated. Each square or triangle equals one trial (*Thomomys*, 18; *Geomys*, 19; *Pappogeomys*, 17).

There was no apparent trend for heavier animals to travel farther or less than lighter animals for all three species in trials run during moist or dry conditions, at 10 to 18° or 21 to 29°C ambient temperature.

Ground-breaking

Despite marked digging adaptations, initial penetration of the surface by the pocket gopher was sometimes difficult in hard, dry soils. One *Geomys bursarius* persisted in clawing at a spot on the ground for seven minutes with its body spinning around in a circle while attempting to start the burrow. In extremely dry, loose soil, pocket gophers sometimes "probed" (abandoned excavations without any 180° turns to push loosened soil) forwards in linear fashion without attaining any depth, producing an open trench. In other instances recorded for all three species of pocket gophers, false starts (excavations abandoned after initial excavation but involving at least one 180° turn) were evident.

Vegetation would occasionally be cut by the incisors and pushed into one of the pouches with the foreclaws by all species, causing minor difficulty in the later execution of 180° turns. All three species were able to twist on the side and back to position the incisors for nipping surface vegetation and digging around rocks and roots. The rigid tail (Fig. 1) used for bracing and orientation of the pocket gopher during excavation assumed a more active role once the front portion of the body became established in the excavation.

The compass direction in which tunnels were initially excavated are represented in Fig. 2. For total trials (56) for all species, tunnels were initiated primarily in easterly (29 per cent) and westerly (17 per cent) directions. For individuals species trials only, *Geomys* (19 trials) did not appear to favor a particular direction, but did have a tendency to dig in southerly (21 per cent), southeasterly (16 per cent), and easterly (26 per cent) directions. *Thomomys* (18 trials) favored a westerly direction 35 per cent of the time (next most frequent was the east with 20 per cent of the trials), whereas *Pappogeomys* (17 trials) strongly favored (41 per cent) an easterly direction with the northeast next most frequent with only 18 per cent of the trials.

Excavation

The high, arching back of the crouching posture as dirt accumulates under the pocket gopher before being kicked away from the burrow entrance with the hind legs is shown in Fig. 1. Crouching enables pocket gophers to balance on hind legs, freeing the front legs for digging at a steeper angle. When the pocket gophers encountered large rocks and roots, excavations normally proceeded in the path of least resistance.

The directions (left, straight, or right) in which tunnels were excavated following ground breaking are indicated in Fig. 3. For individual species trials, *Geomys* (14 trials) favored the left direction 50 per cent of the time compared to only 29 for the straight and 21 for the right directions. *Thomomys* (19 trials), however, overwhelmingly (58 per cent) chose to dig straight in preference to left (21 per cent) or right (21 per cent), as did *Pappogeomys* (40 per cent straight, 27 left, and 33 right). For all trials of each species except *Geomys*, there was no great disparity between the number of left and right choices for tunnel direction.

When the loosened soil was swept under the pocket gophers with the forefeet, digging gradually became more difficult, necessitating a simultaneous kicking of the hind legs. The kicking of *Thomomys bottae* appeared less vigorous (appearing more like a firm shove) than for the other two species, allowing dirt to accumulate behind the animals. This weak kick is unusual in that *Thomomys* had the best jumping capability of the three species, frequently climbing out of cages. In contrast, *Geomys bursarius* kicked dirt 0.3 meters or more beyond the burrow entrance. Propulsive thrusts of the hind legs were more forceful in drier soil conditions.

When the tunnel became too deep to kick dirt out directly, all three species infrequently employed "mule-kicking" (instead of turning, the animal kicks dirt backwards with the hind legs while moving posterior first in inch-worm fashion to the entrance) moving dirt to the tunnel entrance for up to 10 minutes. However, once the soil could no longer be kicked free of the tunnel, pocket gophers normally turned 180° to push dirt to the surface (this response can normally be elicited by putting some loose soil at the back of the pocket gopher as it is excavating). All species normally executed a turn within the first two minutes of excavation.

The direction in which turns were made is presented in Fig. 4. For individual species trials, *Geomys* (8 trials) had the largest difference between

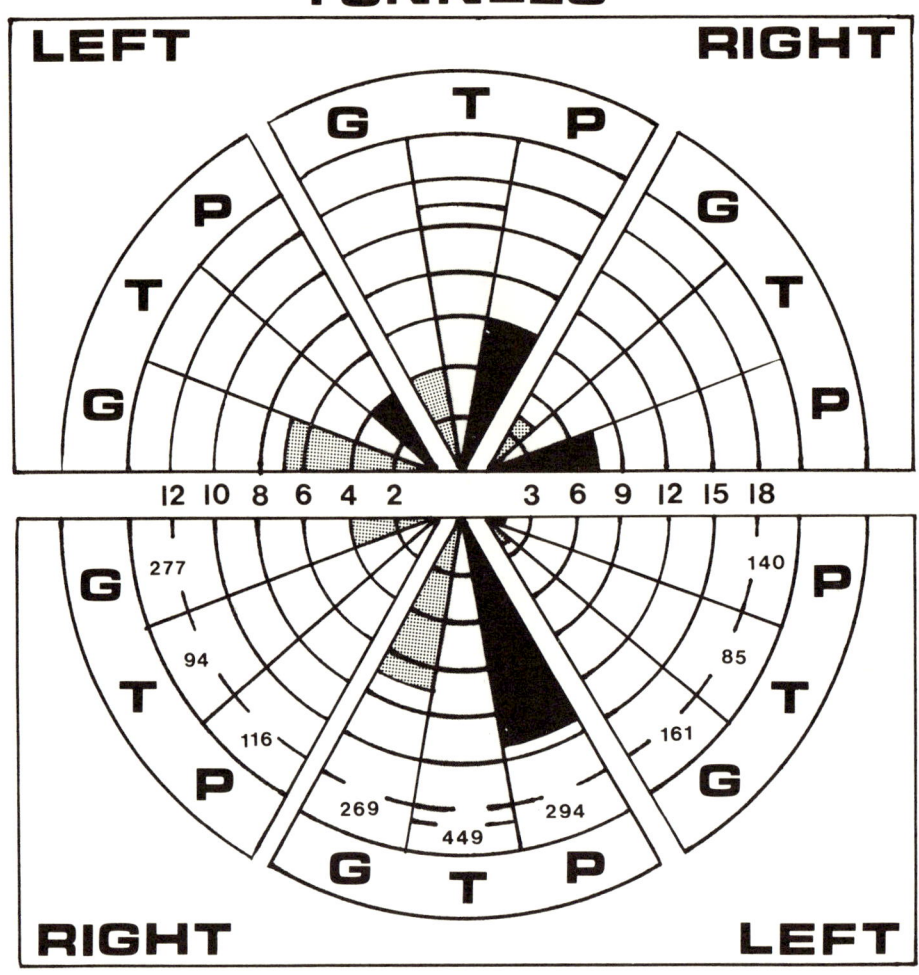

Fig. 3.—Directions taken for constructing the tunnel as opposed to thrusting dirt on the surface. The upper square represents the direction of tunnels excavated by *Geomys* ($N = 14$), *Thomomys* ($N = 19$), and *Pappogeomys* ($N = 15$) following initial penetration of the surface. Each concentric circle is in increments of two excavations. The lower square represents the direction of dirt thrusts onto the surface by *Geomys* ($N = 18$), *Thomomys* ($N = 19$), and *Pappogeomys* ($N = 15$) during mound construction; each concentric circle is in increments of three trials in which a particular direction predominated (numbers within the circles indicate the total number of thrusts of all trials for a particular species). Note that left and right orientation reverses as the pocket gophers turn in the burrow to push the excavated soil.

left (38 per cent) and right (62 per cent) turns, whereas *Thomomys* in trials (53 per cent left, 47 per cent right) and *Pappogeomys* in trials (50 per cent left, 50 per cent right) had progressively less preference for the left or right.

TURNS

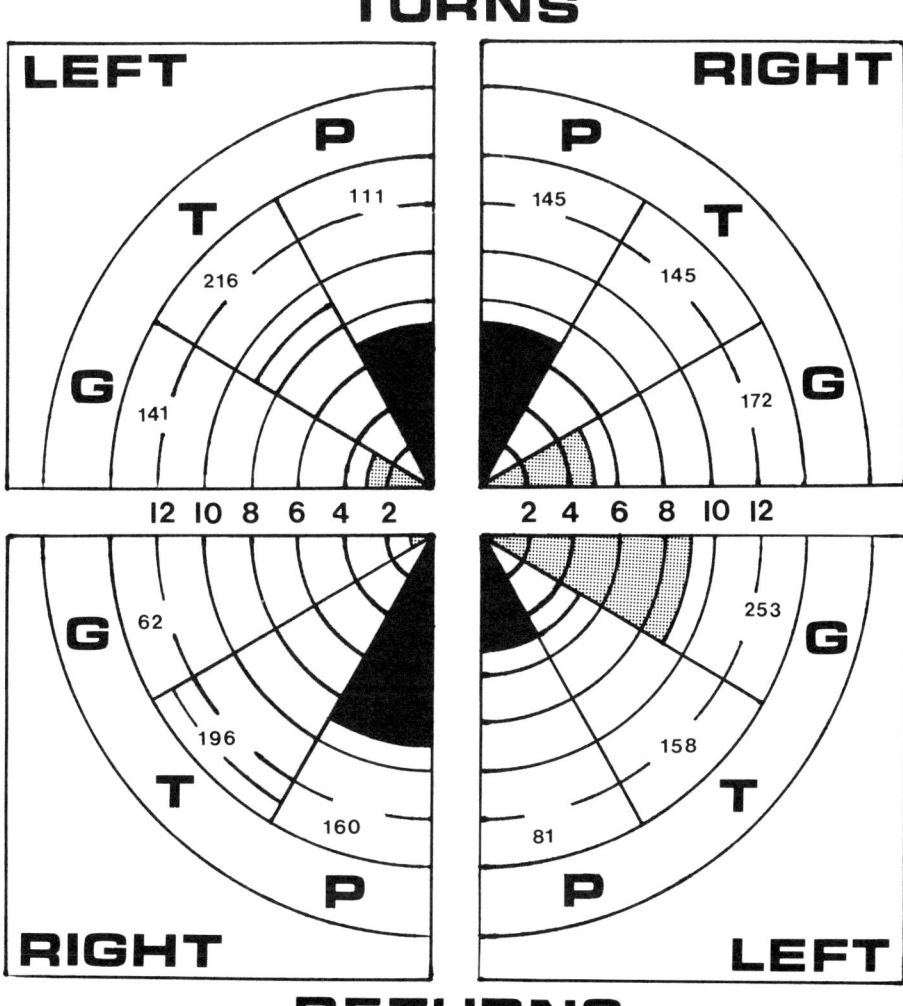

RETURNS

FIG. 4.—About-faces: The top two squares represent the number and direction of left and right turns by *Geomys* (*N* = 8), *Thomomys* (*N* = 17), and *Pappogeomys* (*N* = 14). The lower two squares represent the number and direction of returns by *Geomys* (*N* = 10), *Thomomys* (*N* = 17), and *Pappogeomys* (*N* = 14) during initial mound excavation. Concentric circles represent number of trials in which a particular direction predominated, whereas numbers within the circles indicate the total number of turns or returns of all trials for a particular species. Note that the left and right orientation reverses as the pocket gopher turns in the burrow to push the excavated soil.

When looking at the total number of turns for each species instead of trials, the order of species and magnitude of differences in choice changes to: *Thomomys* (361 turns), 60 per cent left, 40 per cent right; *Pappogeomys* (256 turns), 43 per cent left, 57 per cent right; *Geomys* (313 turns), 45 per cent left, 55 per cent right.

When "scouting," pocket gophers made turns without pushing any soil to the entrance. Scouting was most prevalent in slower excavations where there are longer intervals of time between pushing tailings to the surface.

Moisture affected both the shape of the excavation and the behavior of the excavator. Burrows excavated in hard, dry soil were more irregular in cross section and smaller in diameter than were burrows in moister soils. The angle of the tunnel from the surface in these hardened soil areas was much steeper than the more usual 45° angle for *Geomys bursarius* and *Pappogeomys castanops*, often approaching 70 or 80°. Excavations of *Geomys* and *Pappogeomys* were less steep in moist areas, *Thomomys bottae* never constructed a deep initial tunnel, which conforms with the shallow tunnel in moist soils finding (all 19 trials of *Thomomys* were in moist areas). Temperature had no discernable effect during excavation.

Distribution

Alternating with the excavation phase, the loosened soil is now distributed about the burrow entrance to form the typical earth mound. The first few loads of soil pushed to the surface facilitated additional pushes by flattening surrounding plant growth. Soil scattered about the mouth of the tunnel during kicking become redistributed with the first load pushed onto the surface in normal fashion with the forelegs. All three species nipped off vegetation that occluded the mouth of the burrow or obstructed the passage of tailings to be positioned at the front of the entrance. All species sometimes left a load of dirt at the burrow entrance without further distribution onto the surface ("partial pushes"). Normally, *Pappogeomys castanops* and *Thomomys bottae* were more inclined than was *Geomys bursarius* to extend more than a body length from the burrow opening when thrusting soil onto the surface.

The directions (left, straight, or right) in which thrusts were made are indicated in Fig. 3. For individual species trials, all species predominantly thrust soil onto the surface directly ahead of the tunnel entrance: *Geomys*, 56 per cent (18 trials); *Thomomys*, 100 (19 trials); and *Pappogeomys*, 93 (15 trials). The only disparity between left and right thrusts was a moderate 22 per cent difference (33 right, 11 left) for *Geomys*. However, when looking at the total number of thrusts instead of trials for each species, the picture alters slightly with equal straight (38 per cent) and right (39 per cent) thrusts for *Geomys* (707 total thrusts).

Quantitative data regarding the rate of soil thrusts is enumerated in Table 2. The number of thrusts ranged from 18 to 99 (mean, 44; 19 trials) for *Geomys*; 14 to 59 (mean, 37; 15 trials) for *Pappogeomys*; and 18 to 44 (mean, 26; 19 trials) for *Thomomys*. Time involved in mound building ranged in minutes from 10 to 140 (mean, 51) for *Geomys*, 9 to 57 (mean, 24) for *Pappogeomys*, and 7 to 57 (mean, 26) for *Thomomys*. Rate of soil thrusts in pushes per minute ranged from 0.3 to 1.8 (mean, 1.1) for *Geomys*,

TABLE 2.—Rate of dirt thrusts by mound-building Geomys bursarius, Pappogeomys castanops, and Thomomys bottae under varying conditions of temperature and soil moisture. See Table 1 for explanation of abbreviations.

	Geomys bursarius						Pappogeomys castanops						Thomomys bottae				
Grams/Sex	Dry-moist	°C	Thrusts	Min.	Thrusts/min.	Grams/Sex	Dry-moist	°C	Thrusts	Min.	Thrusts/min.	Grams/Sex	Dry-moist	°C	Thrusts	Min.	Thrusts/min.
75M	M	13	68	45	1.3	218M	M	24	23	13	1.8	59F	M	16	37	30	1.2
75M	M	27	44	40	1.1	218M	M	16	14	10	1.4	60F	M	16	40	57	0.7
75M	M	27	66	63	1.0	283M	D	16	24	10	2.4	60F	M	18	35	24	1.5
75M	M	22	99	68	1.4	283M	M	16	25	18	1.4	60F	M	24	40	27	1.5
105F	M	22	55	35	1.6	310F	M	24	37	38	1.0	60F	M	16	44	39	1.1
173F	M	10	18	10	1.8	310F	M	18	23	9	2.6	69F	M	16	22	8	1.8
173F	D	10	24	62	0.4	310F	M	16	58	26	2.2	69F	M	18	18	28	0.6
173F	M	13	30	20	1.5	310F	M	29	46	28	1.6	69F	M	24	28	56	0.5
180F	D	18	60	77	0.8	310F	D	27	43	57	0.8	75F	M	16	41	33	1.2
183F	D	16	21	13	1.6	316F	M	16	31	25	1.2	87M	M	16	24	10	2.4
183F	M	22	50	52	0.9	320F	M	18	48	23	2.1	88M	M	16	41	12	3.4
223M	D	22	44	33	1.3	320F	M	24	59	28	2.1	88M	M	18	42	20	2.1
223M	D	10	58	81	0.7	320F	M	18	46	45	1.0	88M	M	24	34	45	0.8
223M	M	27	45	44	1.0	353M	M	24	40	18	2.2	88M	M	16	36	18	2.0
228F	D	10	25	45	0.6	353M	M	18	32	12	2.7	94F	M	16	24	18	1.3
269M	M	10	32	32	1.0							94F	M	23	42	24	1.8
269M	D	10	40	140	0.3							94F	M	16	28	33	0.9
269M	M	13	27	24	1.1							98M	M	16	27	7	3.9
269M	D	10	34	55	0.6							98M	M	23	25	12	2.1
X̄ ± SD		44±4.6	51±7.3		1.1±0.1	X̄ ± SD		37±3.6	24±3.6		1.8±0.2	X̄ ± SD		33±1.9	26±3.3		1.6±0.2

0.8 to 2.7 (mean, 1.8) for *Pappogeomys*, and 0.5 to 3.9 (mean, 1.6) for *Thomomys*.

Once excavation commenced, all three species dug rapidly and intensely, disappearing below ground quickly. However, a low rate of pushes per minute may result from a number of delays. Of minor consequence were pauses (termed "freezing") when pocket gophers scanned the surrounding area before or after pushing tailings from the entrance. Wriggling segments of severed earthworms (pushed onto the surface with the loosened soil), coleopterans, and dipterans (which gathered and alighted on the moist, excavated soil) had no noticeably distracting effect on mound-building.

Grooming and eating (Vaughan, 1969), resting, and "forays" had some disruptive influence. Grooming generally lasted only several seconds, involving a few swipes of the foreclaws behind the ears and forwards to the nose. However, in extremely muddy soil, pocket gophers sporadically rested on their hindlimbs for as long as a minute and cleaned the foreclaws and haunches with their incisors. Longer delays of up to 10 minutes occurred when pocket gophers stopped to eat (tooth chattering was clearly audible), or tucked their head downwards in sleeping position with eyes closed. Contrary to resting, one or two individuals of each species abandoned the tunnel and explored the area round the burrow entrance (forays) for distances of up to 10 feet before resuming excavation.

Soil moisture also influenced digging rate. For *Geomys*, the mean for number of thrusts under dry conditions was 42 (8 trials) compared to a comparable mean of 49 (11 trials) under moist conditions. However, the mean time in minutes rose from 39 under moist conditions to 63 under dry conditions, which decreased the rate of soil thrusts from 1.3 to 0.7. Soil moisture was not variable enough during trials for *Pappogeomys* and *Thomomys* for similar comparisons.

Temperature differences under moist conditions (trials under dry conditions were too few for analysis) had little effect on rate of digging in pushes per minute for *Geomys* (1.4 at 10-18°C in 5 trials, to 1.2 at 21-29°C in 6 trials) or *Pappogeomys* (1.6 both at 10-18°C in 8 trials and 21-29°C in 5 trials). Thrusts (35 to 41) and minutes (21 to 25) remained about the same for *Pappogeomys*; during higher temperatures, however, both thrusts (35 to 60) and minutes (26 to 50) increased to give a similar thrust per minute rate as that obtained at lower temperatures. The largest difference for a species in pushes per minute comparing the two different ranges in temperature was for *Thomomys*: 1.4 at 10 to 18°C in 14 trials, decreasing to 1.0 at 21 to 29°C in 5 trials, a result of an increase in the minutes from 24 to 33 (thrusts remaining about the same at 33 and 34).

In alternating the distribution phase of mound building with the excavation phase, the pocket gopher makes a 180° about-face to return to the digging site after distributing tailings on the surface with a soil thrust. These returns to the left or right were usually executed in the easiest and most direct manner. For instance, a pocket gopher pushing soil to the right made right returns of 90° instead of making a 270° return to the left. Sometimes a particular direction of returning remained unaffected by the heaping of soil so that returns were favored in a particular direction (Fig. 4). For individual species trials, *Geomys* (10 trials) overwhelmingly favored left (90 per cent) over right returns whereas *Thomomys* (18 trials) strongly favored the right returns (72 per cent) over left; *Pappogeomys* also favored the right return (64 per cent) over the left. However, when looking at total number of returns for each species instead of trials, the differences between right and left returns decreases somewhat: *Geomys* (315 returns) 80 per cent left to 20 right; *Thomomys* (354 returns) 55 right to 45 left; *Pappogeomys* (241 returns) 66 right to 44 left. Nonetheless, a disparity in left and right returns persists for all three species but is not consistent to the left or right for all three species.

Plugging

The last few loads of excavated soil occlude the tunnel entrance, and several additional loads of soil are positioned at the entrance and packed solid to form the plug. Only in one instance when a cave-in occurred during excavation by *Geomys bursarius* did a pocket gopher begin pushing soil out from another opening before the initial entrance was sealed.

DISCUSSION

Mound-building behavior of *Geomys bursarius*, *Pappogeomys castanops*, and *Thomomys bottae* was essentially similar; likewise, *Geomys bursarius* from western Texas and *G. pinetis* in central Florida (Hickman and Brown, 1973) exhibit many similarities in mound construction and burrow structure (Brown and Hickman, 1973; Hickman, 1977a) despite geographic isolation. Anatomical adaptations to the fossorial habit are similar not only for genera of a family or even order, but also between orders (Shimer, 1903; Ellerman, 1956). Similarity in digging requirements and ecological niche promotes convergence in structure and metabolism (McNab, 1966) which in turn promotes covergence in behavioral repertoire. The adaptive significance of each of the mound-building phases will now be considered.

Prospecting

There are two alternatives for the exposed pocket gopher on the surface: 1) seek refuge by digging immediately; or, 2) investigate the immediate area before digging, which might result in digging being accomplished with a higher degree of efficiency. Contrary to expectations, all three genera spent considerable time (sometimes more than an hour) wandering in random fashion on the surface despite daylight, temperature extremes and danger of predation (all of which are of utmost importance to animals highly adapted to the subterrestrial niche). A previous study on *G. pinetis* (Hickman and Brown, 1973) had indicated a maximum of only 20 minutes before digging commenced in sandy soil in Florida.

Actually, the three species of pocket gophers studied seemed capable of surviving indefinitely on the surface, provided there was presence of adequate cover or darkness (Bryant, 1913), succulent food, and the absence of predators, which would be hindered by defences of the pocket gopher normally employed in burrows (Hickman, 1977*b*). Less than perfect conditions, however, prevailed during testing. Dry soils at 10 to 18°C decreased the meters traveled by 40 per cent and minutes on the surface by 30 per cent for *Geomys*. Temperature increase, even during moist conditions, tended to decrease the meters traveled for all species and minutes on the surface except for *Pappogeomys*; perhaps being the largest species and least able to lose body heat, rather than digging and increasing metabolic heat, the better alternative is to travel slowly on the surface for cover or a suitable digging site. As a result, rate of travel on the surface for *Pappogeomys* decreased by 50 per cent. McNab (1966) has reported on the problems of heat loss in fossorial mammals. Nonetheless, no clear trend was found for heavier animals within the three species to travel farther faster under any of the conditions tested (which might be inconclusive, inasmuch as most tests were made during moist conditions when surface traveling is favorable as the section on barriers later indicates). Other than the physiological advantage of being able to lose heat faster (no tests were run in extreme cold weather), the young of all geomyid species are probably comparably adapted for surface dispersal as the adults inasmuch as fossoral mammals are generally K-selected and unable to sacrifice large numbers of young to hazardous overland travel.

In spite of difficulties, surface travel is an important distributional factor for pocket gophers traversing edaphic barriers such as impenetrable rock strata, water-logged soils, or loose sand, inasmuch as all three species readily crossed asphalt and concrete paved areas with little or no hesitation. Likewise, puddles presented little problem to prospecting gophers, as should be expected, considering their swimming ability (Hickman, 1977*c*). Water, however, might be a psychological instead of physical barrier. Although *Geomys bursarius* readily entered shallow puddles in prospecting, it only reluctantly ventured into deep waters. Flooding of burrows might have great signficance in the range extension of geomyids by causing displacement of pocket gophers from established burrow systems, which would otherwise only be abandoned as a last resort. Spring melt waters in the mountains might initiate exploratory activity in *Thomomys bottae* on a regular seasonal basis. Conversely, the poorer swimming ability of *Pappogeomys castanops* makes the prospect of regular flooding less advantageous, perhaps one explanation why *Thomomys bottae* has been found concentrated in lower lying areas while in competition with the larger *Pappogeomys castanops* (Reichman and Baker, 1972).

Aside from instigating dispersal, flooding also provides conditions condu-cive to dispersal. At times of flooding, atmospheric humidity is high, water in the form of puddles is available for drinking, and the moist soil is loose, shortening excavation time. Such factors are of additional significance in summer when substrate temperatures during unflooded conditions are likely to be higher than ambient temperatures. During this study, one *Geomys bursarius* was nearly overcome by heat and would have perished (as did a prospecting *Pappogeomys castanops*) had it not reached the shade of a tree. A study of mound-building during the night as opposed to the day would be most interesting, inasmuch as activity patterns of fossorial mammals generally extend over 24 hours with only intermittent rests (Hickman, 1980).

Ground-breaking

False starts (unfinished excavations) by *Pappogeomys castanops*, *Thomo-mys bottae*, and *Geomys bursarius* noted in this study, and for *Geomys pinetis* in Florida (Hickman and Brown, 1973) suggest that geomyids may be somewhat selective in their excavation site and are not committed immediately to seeking shelter underground. One might suspect that even a temporary burrow might be hurridly excavated and occupied until night when lower temperatures and darkness would be more conducive to overland travel. On the negative side of the ledger, predation pressure might be greater at night, particularly in the form of owls, the pellets of which sometimes yield great numbers of pocket gopher skulls (Bird, 1929; and many others).

The presence of ground cover was not a prerequisite for digging, although many excavations began at the base of grass clumps. Plants nipped off and stuffed in the pouches during ground-breaking permitted pocket gophers to eat while resting from digging or after plugging when the only food available might require a considerable amount of additional digging; also, energy reserves might be low following excavation. Turning and twisting on the side and back better to position the incisors for gnawing through vegetation and around rocks has previously been reported by Hickman and Brown (1973) for *Geomys pinetis* and Wight (1918) for *Thomomys bulbivorous*. Use of the tail in orientation and bracing while gnawing (Hickman and Brown, 1973) was most characteristic of *Geomys bursarius*, whereas such use of the tail by *Pappogeomys castanops* and *Thomomys bottae* was more restricted to excavation in sun-baked and rocky soils. A comparative study of the use of the tail by geomyids and shorter-tailed fossorial mammals during excavation would be of interest (Hickman, 1979*b*).

Unexpected was the preference for particular compass directions when initiating excavations: *Pappogeomys* dug eastwards 41 per cent of the trials, *Thomomys* westward 35 per cent of the trials, and *Geomys* eastward 26 per cent of the trials. Such a preference for direction might be adaptive for species with only a few burrow entrances, which are left unplugged (especially from a particular direction). However, except for a mountainous habitat for *Thomomys*, none of those factors applies to pocket gophers. The answer is more probably due to proximal causes at the time of excavation. Because the sun always traces an east to west path, perhaps pocket gophers normally face away from the sun on bright days so as not to be blinded while scouting the area for predators during excavation. Moreover, since the eyes are laterally located on the head, even facing towards the sun would not appear to be detrimental. A simpler alternative might be that direct sun is irritating, particularly to eyes accustomed to darkness.

Excavation

Once digging commenced, *Thomomys* dug tunnels straight 58 per cent of the trials and *Pappogeomys* 40 per cent of the trials. There was no great disparity between left and right tunnel directions for *Thomomys* or *Pappogeomys*, but *Geomys* favored the left over straight or right 50 per cent of the trials. Straight tunnels might demand less energy when the pocket gopher pushes dirt to the surface; pushing around a corner rather than straight ahead probably involves a greater effort, at least under moist and therefore heavier soil conditions prevalent under trials for *Thomomys* and *Pappogeomys*. Under extreme dry conditions, *Geomys bursarius* and *Pappogeomys castanops* modified mound-building behavior by digging at a steep angle through the hard surface crust to the soft, moist soils below with deviations to the right or left only upon reaching the softer underlying soil. At other times, veering to the right or left could have been an attempt to bypass endurant soils, which were more prevalent during trials for *Geomys*. The reason for the left direction being preferable to the right for *Geomys* is not easily determined.

Although mound-building in all three species was sufficiently flexible to allow changes in the direction of excavation, flexibility of the behavioral repertoire raises the question of why pocket gophers chew through underground cables (Howard, 1953a) instead of excavating around them. Perhaps cables make excellent objects on which to sharpen and wear down the rapidly growing nails (Howard, 1953b; Kennerly, 1958) and incisors (Howard and Smith, 1952; Miller, 1958). More importantly, it is easier to attempt removal of obstructions from the tunnelway underground where work may proceed for long periods without problems of desiccation or predation.

Elimination of initial turns and use of mule-kicking during dry conditions enables the animals to reach a suitable depth faster than if they had to turn to push soil with the forefeet. The fact that kicks were the least propulsive for *Thomomys* despite being the best jumper of the three species can best be explained by soil conditions. Moist soil (the rule of all trials of *Thomomys* in this study) tends to cling and is less likely to fall back into the tunnel and disrupt digging. Additional testing is needed for *Thomomys* under dry conditions.

Turns are similar in all three species, although the about-faces of *Geomys bursarius* described by Breckenridge (1929) and Downhower and Hall (1966) were sometimes rendered more difficult by plant material carried in the pouches. Differences in percentages of left and right turns for total turns of each species were somewhat disparate: a 20 per cent difference for *Thomomys*, and a 14 per cent difference for *Pappogeomys* and *Geomys*; however, there is no apparent selective advantage for a pocket gopher to favor either a left or right turn.

Regardless of direction, pocket gophers normally do turn to push soil, which is not the case for other fossorial mammals such as *Ctenomys* (Pearson, 1959). Predators are more likely to grab the exposed hindquarters than confront the claws, incisors, and hissing of an alerted pocket gopher. Also, when digging without turns during the mule-kicking, soil is not arranged in typical fashion about the burrow entrance, which normally allows for construction of the typical conical mound and concealment of the plug. It is the gradual angle of the tunnel downwards which allows pocket gophers to turn; otherwise the soil would have to be regathered with each turn before pushing soil to the surface. *Pappogeomys castanops*, *Thomomys bottae*, and *Geomys bursarius* never transported tailing in the cheek pouches in contrast to early reports of such behavior in *Geomys bursarius* by Parvin (1855) and Webster (1897). Merriam (1895), Bailey (1895), and all later investigators have never recorded soil being carried by pocket gophers in the cheek pouches.

Distribution

The manner of distributing dirt on the surface was not qualitatively different for the three species. None of the three species consistently paused prior to pushing soil from the burrow, perhaps because of the proximity of thrusts to the burrow made pauses to scan the surface for predators unnecessary. Futhermore, the pauses that did occur were not accompanied by a reorganization of the soil load as reported for *Geomys bursarius* by Breckenridge (1929). All three species infrequently pushed tailings more than a body length from the burrow entrance, in contrast to *Geomys pinetis* in Florida (Hickman and Brown, 1973) where the long ramp length allowed for the total number of soil thrusts to total as high as 102. Flinging the soil from the exit with a flip of the nose as reported for *Geomys bursarius* (Scheffer, 1910) was not noted. However, all three species were able to thrust the tailings out in a spray up to 10 centimeters around the burrow entrance, with the last propulsive push with the forepaws launched from the hunched, inchworm posture (Hickman and Brown, 1973). The spray of soil is generated in much the same manner as one splashes water by rapidly thrusting the palm forward through the surface of water.

All three species favored straight thrusts by at least 50 per cent of the trials; *Geomys* was the only species with a preference for right (33 per cent) over left (11 per cent). Because nonstraight thrusts probably require more energy in turning and stretching, straight thrusts should be expected to be prevalent; however, the secondary preference for left pushes by *Geomys* does not have any obvious adaptive value.

The large range in number, minutes, and rate of soil thrusts reflects the large range in environmental conditions under which trials were conducted. In moist soil at 10 to 18°C, however, the mean for the number of thrusts performed per excavation was close: *Geomys*, 35; *Pappogeomys*, 35; and *Thomomys*, 33. With similar environmental conditions, physical appearance, and behavioral patterns, convergence of digging ability may result. At the same temperature range (10 to 18°C) but under dry conditions, *Geomys* still utilized approximately the same number of thrusts (mean, 37), but required almost twice as long (a mean of 26 minutes increased to 68), which more than halved the thrust rate to 0.5. Prospecting for a suitable site can therefore result in a considerable savings of energy in relation to a fixed amount of work. At higher temperatures (21 to 29°C), however, the mean number of thrusts varied greatly between species: (*Thomomys*, 34; *Pappogeomys*, 25; *Geomys* 60). Size and physiological differences in ability to lose heat may have in some cases encouraged early plugging resulting in fewer thrusts; or in other cases, less soil may have been excavated per load in an attempt to plug the entrance quickly, resulting in a larger number of thrusts. Times of excavation also rose for all species at higher (21 to 29°C) temperatures (perhaps due to more delays for cooling body temperature), so that the mean for thrust rates (thrusts per minute) remained essentially the same for *Geomys* (1.3-1.2) and *Pappogeomys* (1.7-1.6). Temperature increase affected *Thomomys* most (a species normally adaptive to colder climates) with the mean for minutes increasing from 24 to 33, dropping the mean for thrust rate from 1.4 to 1.0

Left and right returns were largely disparate for all species (*Geomys*, 90 per cent left; *Thomomys*, 72 right; *Pappogeomys*, 64 right) but was not consistent for all three species. Detection of tunnels is rendered much more difficult by disparities in the number of left and right thrusts (Hickman and Brown, 1973). Work patterns are established whereby tailings are pushed consistently in one direction during particular periods of mound buidling. Sequences of thrusts, turns, and returns can be altered by the accumulation of soil at a particular position at the mouth of the burrow or the change in body posture necessitated by a change in direction of excavation. *Geomys*, it may be recalled, appeared to favor a particular left or right direction for tunnels, turns, thrusts, and returns, whereas the only instances of preferential directionality to the left or right by *Thomomys* and *Pappogeomys* were during returns. Additional testing under more standardized conditions is needed to clarify the significance of directionality in all phases of mound building in *Geomys*.

Plugging

Packing soil into the opening of the entrance is important as protection for the pocket gopher against flooding, extreme temperatures, and other organisms (Hickman, 1977b), so that burrows were in no instances left unplugged.

Mound Building

Why do pocket gophers organize the excavated soil into surface mounds? Fossorial insectivores such as the talpids and chrysochlorids are able to burrow just below the surface and push soil upwards without the formation of mounds; however, mounds are usually formed when excavating deeper tunnels. Fossorial rodents such as ctenomyids, spalacids, and geomyids on the other hand, always construct surface mounds as a means of ridding their tunnels of excavated materials. The hind feet have become somewhat adapted in geomyids for pushing soil, but not to the extent of *Ctenomys*, which normally utilizes the alternative strategy of pushing soil to the surface with the hind legs and therefore does not turn. Perhaps early in evolutionary history when geomyids were not as fossorial in habit, turns developed which allowed pocket gophers to scan the surface before pushing dirt out of the burrow rather than depending solely on kicking the loose excavated soil haphazardly onto the surface with hind quarters exposed. The primitive burrows of geomyids might have had a heap of soil develop as animals positioned excavated soil to the sides and front of the entrance while at the same time remaining within the security of the burrow. The extra expenditure of energy in pushing the soil flat to make the tell-tale sentinel of soil less conspicuous would not have been warranted because the entrance to the tunnel might have been made even more conspicuous from a distance by a large bare patch of soil. The habit of plugging the open entrance of the volcanolike mound might have developed slowly with other changes associated with increasing aridity (Russell, 1968) and became equally important as a deterrent to other would-be symbionts and predators as well as to the elements. Mounds can even now be seen with the plug very clearly demarcated, but most mounds are constructed so that the heaped up soil conceals the exact location and direction of the plugged tunnel.

Why do the three genera construct mounds in similar fashion? There has been a limited amount of time for behavioral divergence since the geomyid Pliocene-Pleistocene radiations. In fact, many of the pronounced fossorial adaptations of geomyids arose in the latter part of the Miocene and early Pliocene before the geomyid tribes Thomomyini (*Thomomys*) and Geomyini diverged (Russell, 1968). The basic similarity of behavioral repertoire of those two tribes likely is associated with similar appearance inasmuch as there is usually a close correspondence between morphology and behavior (Colbert, 1969). The ability to create an environment to suit humidity and temperature requirements and diminished contact with the large and diverse surface fauna might have placed the geomyids in such a static situation that adaptive pressures favoring divergence have and continue to be few. It is because of such similarities that competitive exclusion is usually the rule not only for geomyids, but fossorial mammals in general (Nevo, 1979).

As a final note, this study was intended as a preliminary step in determing the behavior repertoire and evolutionary trends of three genera of geomyids involved in mound building. To term certain findings as significant because of a statistical test when trials were conducted under unstandardized conditions of animal size and approximate age, soil hardness and temperature, amount of moisture, type and density of vegetation, light intensity and direction, time of testing, and other variables did not seem warranted. Additional studies on the mound-building behavior of other genera of geomyides (*Orthogeomys* and *Zygogeomys*) and other species are needed to provide a broader base for comparison. In particular, knowledge of mound building in tropical environments might clarify the influence of moisture, soil types, and temperature on behavioral patterns involved in prospecting and mound construction. Rather than make premature statements on significance, enclosures constructed outside and in environmental chambers where various parameters involved in mound building could be isolated and their relative importance evaluated should be undertaken. Perhaps most important of all, comparison with other families of fossorial mammals should reveal and clarify alternative strategies of excavation (perhaps including shorter or longer prospecting times, methods of initial ground breaking, vertical excavations, pushing loosened soil with the hind feet, favoring particular direction during phases of excavation, to cite a few). Only then might the most prominent convergent construction of eutherian fossorial mammals and their associated behavior be fully appreciated—the loose conical heap of excavated soil, the surface mound.

ACKNOWLEDGMENTS

This research formed part of my doctoral dissertation at Texas Tech University under the direction of Robert L. Packard. Helpful comments were received from Dilford C. Carter, the editors and reviewers. Live traps were kindly made available by Robert J. Baker. The typist, Susie Hickman, labored under the slings and arrows of outrageous handwriting. Financial support from grants-in-aid of research from the Society of Sigma Xi and the Theodore Roosevelt Memorial Fund of the American Museum of Natural History is gratefully acknowledged.

LITERATURE CITED

BAILEY, V. 1895. Pocket gophers of the United States. Bull. U. S. Dept. Agr. Div. Ornith. and Mamm., 5:1-47.

BAKER, R. J., AND S. L. WILLIAMS. 1972. A live trap for pocket gophers. J. Wildlife Mgt., 36:1320-1322.

BIRD, R.   1929.   The great horned owl in Manitoba.   Candian Field Nat., 43:79-83.

BRECKENRIDGE, W. J.   1929.   Actions of the pocket gopher (*Geomys bursarius*).   J. Mamm., 10:336-339.

BROWN, L. N., AND G. C. HICKMAN.   1973.   Tunnel system structure of the southeastern pocket gopher.   Florida Scientist, 36:97-103.

BRYANT, H. C.   1913.   Nocturnal wanderings of the California pocket gopher.   Univ. California Publ. Zool., 12:25-29.

COLBERT, E. H.   1969.   Morphology and behavior.   Pp. 27-47, *in* Behavior and evolution (A. Roe and G. G. Simpson, eds.).   Yale Univ. Press, New Haven, 557 pp.

DOWNHOWER, J. F., AND E. R. HALL.   1966.   The pocket gopher in Kansas.   Misc. Publ. Mus. Nat. Hist., Univ. Kansas, 44:1-32.

ELLERMAN, J. R.   1956.   The subterranean mammals of the world.   Trans. Roy. Soc. S. Africa, 35:11-20.

HERMANN, J. A.   1950.   The mammals of the Stockton Plateau of northeastern Terrell County, Texas.   Texas J. Sci., 2:368-393.

HICKMAN, G. C.   1977a.   Burrow system structure of *Pappogeomys castanops* (Geomyidae) in Lubbock County, Texas.   Amer. Midland Nat., 97:50-58.

———— .   1977b.   Geomyid interaction in burrow systems.   Texas J. Sci., 29:235-244.

———— .   1977c.   Swimming behavior in representative species of the three genera of North American geomyids.   Southwestern Nat., 21:531-538.

———— .   1978.   A transparent burrow system for the study of fossorial mammals.   Acta Theriologica, 23:443-445.

———— .   1979a.   Burrow system structure of the bathyergid *Cryptomys hottentotus* in Natal, South Africa.   Z. Säugetierk., 44:153-162.

———— .   1979b.   The mammalian tail: a review of functions.   Mammal Rev., 9:143-157.

———— .   1980.   Locomotory activity of captive *Cryptomys hottentotus* (Bathyergidae), a fossorial rodent.   J. Zool., London, 192:225-235.

HICKMAN, G. C., AND L. N. BROWN.   1973a.   Mound-building behavior of the southeastern pocket gopher (*Geomys pinetis*).   J. Mamm., 54:786-790.

———— .   1973b.   Pattern and rate of mound production in the southeastern pocket gopher (*Geomys pinetis*).   J. Mamm., 54:971-975.

HOWARD, W. E.   1953a.   Tests of pocket gophers gnawing electric cables.   J. Wildlife Mgt., 17:296-300.

———— .   1953b.   Growth rate of nails on adult pocket gophers.   J. Mamm., 34:394-396.

HOWARD, W. E., AND M. E. SMITH.   1952.   Rate of extrusive growth of incisors of pocket gophers.   J. Mamm., 33:485-487.

KENNERLY, T. E., JR.   1958.   Comparisons of morphology and life history of two species of pocket gophers.   Texas J. Sci., 10:133-146.

MERRIAM, C. H.   1895.   Monographic revision of the pocket gophers, family Geomyidae (exclusive of the species of *Thomomys*).   North Amer. Fauna, 8:1-258.

MCNAB, B. K.   1966.   The metabolism of fossorial rodents: a study of convergence.   Ecology, 47:712-733.

MILLER, R. S.   1958.   Rate of incisor growth in the mountain pocket gopher.   J. Mamm., 39:380-385.

MIELKE, H. W.   1977.   Mound building by pocket gophers (Geomyidae): their impact on soils and vegetation in North America.   J. Biogeogr., 4:171-180.

NEVO, E.   1979.   Adaptive convergence and divergence of subterranean mammals.   Ann. Rev. Ecol. Syst., 10:260-308.

PARVIN, J. B.   1855.   On the habits of the gopher in Illinois (*Geomys bursarius*).   Annual Rep. Smithsonian Inst. for 1854, 1855:293-294.

PEARSON, O. P.   1959.   Biology of the subterranean rodents, Ctenomys, in Perú.   Mem. Mus. His. Nat. "Javier Prado," 9:1-56.

REICHMANN, O. J., AND R. J. BAKER. 1972. Distribution and movements of two species of pocket gophers (Geomyidae) in an area of sympatry in the Davis Mountains, Texas. J. Mamm., 53:21-33.

RUSSELL, R. J. 1968. Evolution and classification of the pocket gophers of the subfamily Geomyinae. Univ. Kansas Pub. Mus. Natur. Hist., 16:473-579.

SCHEFFER, T. H. 1910. The pocket gopher. Kansas State Agr. Coll. Exp. Sta. Bull., 172:1-39.

SETON, E. T. 1929. Lives of game animals. Doubleday, Doran and Company, New York. Vol. 4:395-418.

SHIMER, H. W. 1903. Adaptations to aquatic, arboreal, fossorial and cursorial habits in mammals. III. Fossorial adaptations. Amer. Nat., 37:819-825.

VAUGHAN, T. A. 1966. Food-handling and grooming behaviors in the plains pocket gopher. J. Mamm., 47:132-133.

WEBSTER, C. L. 1897. The pocket gopher, or pouched gopher (Geomys bursarius). Amer. Nat., 31:114-120.

WIGHT, H. M. 1918. The life history and control of the pocket gopher in Williamette Valley. Oregon Agr. Exp. Sta. Bull., 153:1-55.

WIGHT, H. M. 1942. The Williamette Valley gopher. Murrelet, 23:6-8.

ANALYSIS OF BEHAVIORAL PATTERNS IN POPULATIONS OF SILKY POCKET MICE, GENUS PEROGNATHUS (RODENTIA: HETEROMYIDAE)

ROBERT E. MARTIN

Interest in the specific status of two similar-sized species of pocket mice, *Perognathus flavus* and *P. merriami*, and that of a larger species, *P. flavescens*, included for comparison, prompted a quantative analysis of their behavior and species preferences (Martin, 1974). These rodents are granivorous (Forbes, 1962; Chapman and Packard, 1974), generally inhabiting arid, semiarid, and mesic grasslands of central México, and portions of the Southwest and the Great Plains of the United States (Fig. 1). The smaller species, *P. flavus* and *P. merriami*, resemble one another morphologically (Williams, 1971) and karyologically (Patton, 1967; Williams, 1978). Based on those similarities, Williams (1971) placed both taxa in the same species group, and Wilson (1973) formally proposed that *P. flavus* and *P. merriami* be considered conspecific under the earliest name, *P. flavus*. Hall (1981) accepted that arrangement but suggested that further study of Chihuahuan material would prove to be rewarding for confirmation of that conclusion. In the present study, the species names *P. flavus* and *P. merriami* are used in presenting and discussing these behavioral data.

Osgood (1900) regarded *P. flavus* and *P. merriami* as separate species, based on a combination of characters and their sympatric occurrence at Eddy (=Carlsbad), New Mexico, the type locality of *P. merriami gilvus*. Wilson (1973) concluded that *P. merriami gilvus*, which occupies a geographically intermediate position between *P. flavus* and *P. merriami*, has morphological features intermediate between these two species. Davis (1974) continued to regard these two forms as distinct species, but Schmidly (1977) independently concluded that Wilson (1973) was correct in recognizing only a single species, *P. flavus*. Data from species preference experiments (Martin, 1977) also supported Wilson's (1973) contention. Williams (1978) later suggested that "The broad sympatric zone of *P. flavus* and *P. merriami* appears to be a broad zone of hybridization of the two forms."

The behavioral patterns of three species of pocket mice are presented in this paper. These data resulted from encounter experiments staged to answer the following questions: 1) Are the quantitative and qualitative behavior patterns exhibited by the three species sufficiently distinct to warrant the retention of these taxa as separate species? 2) Is there any evidence of character divergence or character convergence in behavior patterns in areas of sympatry? 3) What characters, or suites of characters, are most useful in analyzing behavioral patterns in these species?

MATERIALS AND METHODS

Subjects

Specimens were captured with either modified Gen traps (2.7 by 18 centimeters) or collapsible Sherman traps (8 by 9 by 23 centimeters), baited with mixed birdseed, from six localities. In the list below, the numbers in parentheses represent the number of mice held in captivity. Only locality names in italics are referred to in the Results and Discussion.

Perognathus flavus.—ARIZONA: Cochise County: 2.4 km. W, 1.6 to 3.2 km. N *Rodeo* (48 males, 40 females). NEW MEXICO: Doña Ana County: *Jornada* Experimental Range: 10 km. S, 0 to 1.6 km. W Headquarters (28 males, 25 females).

Perognathus merriami. TEXAS: Winkler County: 2 km. N, 5 km. W *Wink* (18 males, 35 females); Zavala County: 16 km. S, 6.4 km. W *Uvalde* (30 males, 35 females).

Perognathus flavescens. NEBRASKA: Kearney County, 8 km. S *Kearney* (35 males, 30 females). TEXAS: Winkler County: 3.5 km. N, 10 km. E *Wink* (20 males, 10 females).

Housing and Experimental Apparatus

Mice were housed individually in 3.8-liter glass jars with a sand substrate. Mixed birdseed and commercial guinea pig chow were supplied *ad libitum* and lettuce twice weekly. Temperature and relative humidity averaged $25 \pm 2°C$, and 40 to 50 per cent, respectively. Flourescent illumination was on a 14 light : 10 dark cycle. All experimental trials were conducted inside a glass-fronted observation box within an experimental room. The box was divided into left and right arenas, each 0.3275 square meters, by a guillotine-type partition. A layer of sand, approximately 5 centimeters deep, filled the floor of the arena. Illumination during trials was provided by a 40-watt red flourescent bulb (Westinghouse, F4OR).

To record behavior during the trials, a keyboard was coupled to a 20-channel event recorder (Esterline Angus, A620X). All recordings were made at a chart speed of 7.62 centimeters per minute. Each channel was used to record one type of behavior but additional variables were coded by selecting appropriate pairs of keys.

Experimental Procedure

Encounter trials in a neutral arena were staged to investigate the behavioral patterns. Twenty-six sets of male and female interactions, representing 208 trials (eight trials per set) were tested. To initiate a 10-minute trial (in preliminary observations, an optimal trial length was determined by plotting cumulative number of variables onto time), the sand substrate was stirred thoroughly, smoothed, and the arena divided into two portions by a partition. Two subjects (Ss), male and female, were then placed individually into each portion of the arena. After one to two minutes of acclimation, the partition was raised and the behavior of Ss recorded. The mice were then removed, the substrate stirred, and new Ss introduced.

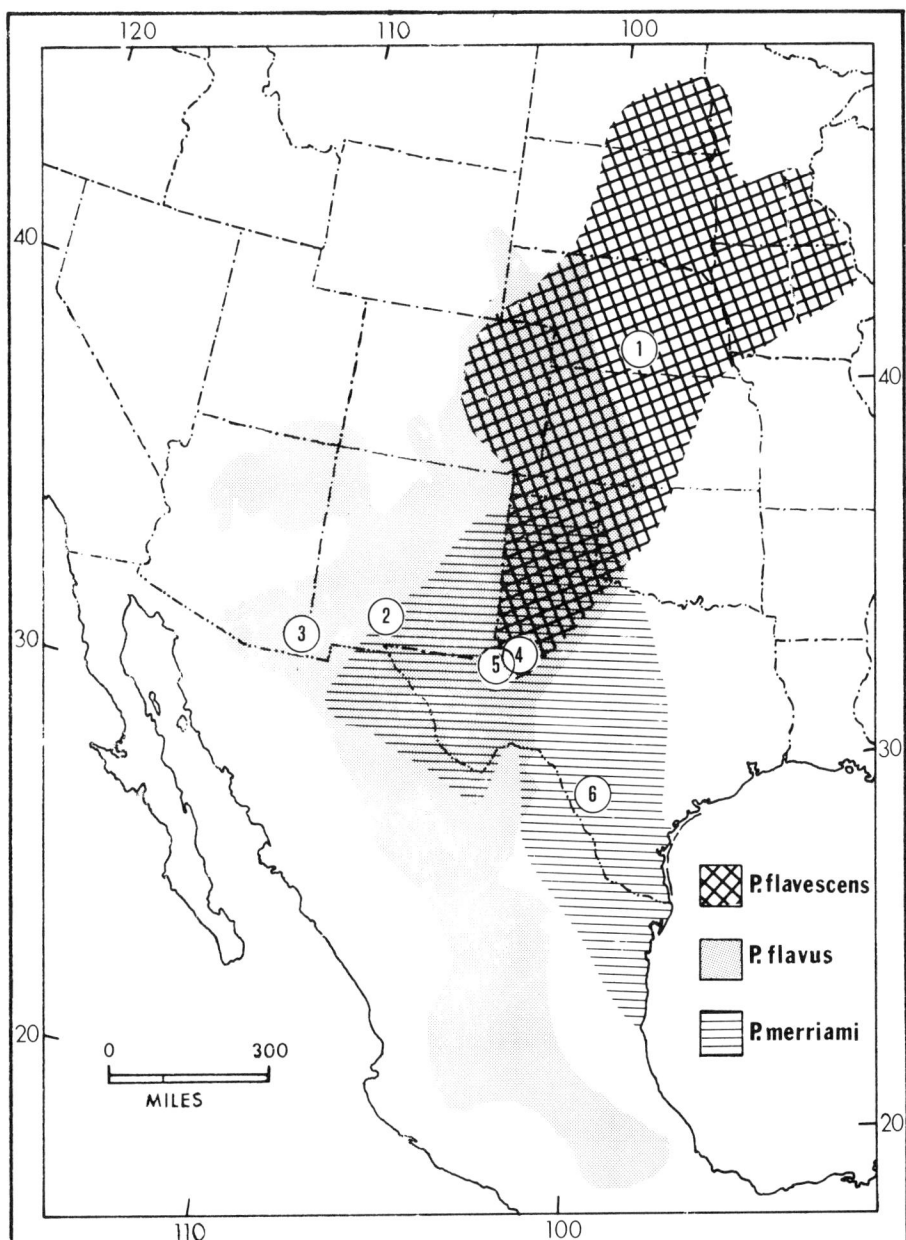

FIG. 1.—Geographic ranges and sites of collection for three species of silky pocket mice, genus *Perognathus*. Ranges based on Hall and Kelson (1959) and localities indentified as follows (details in text): Kearney, 1; Jornada, 2; Rodeo, 3: E of Wink, 4; W of Wink, 5; Uvalde, 6.

TABLE 1.—*List of behavioral variables for the encounter experiments. Abbreviated names are those used in the text and figures. The variable "no change" was reported only in AB (male followed by female behavior) and BA (female followed by male behavior) interactions. Terminology based on Eisenberg (1963a) except as noted otherwise in text.*

Behavioral variable	Abbreviation
Washing and other comfort movements	WASH
Sandbathing (side or ventral)	SNBH
Marking and perianal drag	MARK
Approach (including slow approach)	APPR
Naso-nasal and naso-anal investigation	NNAL
Flight	FLGH
Attack or escape leap	ATES
Withdrawal	WTDL
Upright (types I and II combined)	UPRG
Digging and Kicking back	DGKB
Head over and head under	HOUN
Circling	CRCL
Chase	CHAS
Lock fight	LCFG
Sparring	SPAR
Defeat	DFRT
*Sidling	SDLG
Sand flicking	SDFL
Following or driving (by male)	FLDR
Grooming (another individual)	GMPT
Mounting or riding (by male)	MTRD
Thrusting or intromission (by male)	THIR
Inciting (by female)	INCT
Lordosis (by female)	LRDS
No change observed	NOCH

*never occurred.

Behavioral variables recorded were based on those described by Eisenberg (1963a), with these simplifications: Eisenberg's ventral, left and right sandbathing were recorded as sandbathing; upright 1 and 2 were recorded as upright; slow approach was expanded to include all approaches. A total of 24 different behavioral variables (21 for females, 23 for males) were recorded (Table 1). In this study, an event was defined as a discrete instantaneous act (for example, sand flick) or the start of an extended behavioral sequence of less than 10 seconds duration (for example, digging).

Statistical Analyses

Events (acts) were analyzed to permit intrasubject and intersubject comparisons. All events were transcribed in pairs of sequences to show preceeding and following acts and later transferred to data processing cards.

Original data cards served as sources for 1) summarizing the total number of events per behavioral variable; 2) listing the frequency of events for each behavioral variable with respect to all behavioral variables; 3) constructing a sequential contiguity matrix; and 4) providing a data source deck for use in multivariate analyses (Martin, 1974). A Fortran-IV program, written by William B. Wyatt, was used to obtain values for the first three items. Multivariate stepwise discriminant analysis (BMDO7M), using some behavioral variables, was used to test behavioral differences between species and localities (program described in Dixon, 1971). These analyses and those described below were performed at Computer Services of Texas Tech University on an IBM 370-145.

In the present study, tabulations were made of the occurrence of behavior events for each variable. Such tabulations permit testing the null hypothesis of independence of monad events (Altmann, 1965). Relative frequencies of events (=estimated probabilities of monad events) for each variable with respect to occurrence for all variables were then calculated (converted to percentages for presentation in tables and figures). These relations, summarized mathematically, follow:

$p = (k)/N$ = relative frequency of events for k_{th} variable

  where N = total number of events for all variables

n = observed number of events

k = variable

$n(k)$ = observed frequency of k_{th} variable

Sequence patterns (and associated probabilities) were obtained by tabulating in a contiguity matrix the number of events of a particular variable that followed (or preceeded) another variable (see Martin, 1974). These event "dyads" permit testing the null hypothesis that occurrence of a behavioral variable event is dependent on the immediately preceding event but on none before that (Altmann, 1965). Tabulation of third order (triads), fourth order (quadrads), and higher orders of sequences are possible (see Altmann, 1965; Hazlett and Bossert, 1965). With each higher order, there is a severe decrease in the number of occurrences that fit the stated sequences and a resultant loss of information due to diminished sample sizes. Consequently, only dyads were analyzed in the present study. Conditional probabilities for the occurrence of particular dyads were calculated according to the method of Nelson (1964), as follows:

$P_{jk} = n_1 (j, k)/n_2$, where n_1 = number of dyads of j_{th} variable

  followed by k_{th} variable

n_2 = row sum of occurrences for j_{th} variable

P_{jk} = joint or conditional probability for occurrence of dyad

  j,k. Also designated p (j, k).

Conditional probabilities can be presented in tabular form (Altmann, 1965) or utilized in flow charts (Myrberg, 1972). Using the terminology of Hazlett and Bossert (1965), dyads occurring more frequently than expected were called "directive" and those occurring less frequently than expected were called "inhibitive" to simplify presentation of data and facilitate discussion.

Chi-square tests were used to analyze deviations of observed occurrences of dyads from expected values. Nelson (1964) and other workers noted that lack of independence between behavioral variables is characteristic of many behavioral patterns. Although variable independence is a requisite for making probability statements in chi-square analyses, this technique does provide a measure of deviations from expected values. In the present study, chi-square values for each behavioral dyad were calculated for all cells with expected values>3.5.

Stepwise multivariate-discriminate analysis was used on some count data (monad frequencies of 3 and 4-group comparisons) to reduce dimensions and aid in determining the importance of particular variables for separating groups. Additional details on the use of these analyses can be found in Pimentel and Frey (1978). Recent uses of these analyses in behavioral research include Frey and Miller (1972), Martin (1974), Mykytowycz and Hesterman (1975), Bekoff *et al.* (1975), Aspey (1977), and Pimental and Frey (1978).

RESULTS

Male and Female Behavioral Patterns

Monad frequencies were tablulated for all behavioral varibles that occurred at least once in one or more encounters. Those data are summarized in Martin (1974).

In homospecific encounters, male and female behaviors were similar. Inspection revealed some sexual differences in the frequencies of digging-kick back (DGKB) and sandbathing (SNBH). Within sympatric populations, female *Perognathus flavus* and *P. flavescens* sandbathed significantly more ($P<0.01$) than did sympatric males. There were no significant differences in SNBH ($P>0.05$) between male and female *P. merriami*. In six allopatric, homospecific comparisons, significant differences ($P<0.05$ and $P<0.01$) were observed in the frequencies of SNBH in both tests involving *P. merriami* and in one test involving *P. flavescens*. Significant differences ($P<0.01$ and $P<0.05$) also were apparent between males and females in the frequencies of DGKB in three of six homospecific, sympatric comparisons and three of six homospecific, allopatric comparisons. No trend was evident in those data since some tests revealed both high and low values for males.

In heterospecific encounters, inspection of monad frequencies revealed significant sexual differences in SNBH, DGKB, and washing (WASH) behavior. *Perognathus merriami* and *P. flavescens* males had significantly smaller frequencies ($P<0.01$) of SNBH compared to their female "opponents" (Table 2) in seven of 11 encounters. Frequencies of SNBH were not significantly different between males and females in encounters where *P. flavus* was paired with heterospecifics and in some *P. merriami* and *P. flavescens* encounters (Table 2). No clear trends were evident in the male and female monad frequencies of DGKB. Of 14 heterospecific encounters

TABLE 2.—*Monad frequencies of sandbathing (SNBH) and digging-kickback (DGKB) in heterospecific encounters of* Perognathus. N *is the total number of events.*

Species	N		SNBH		DGKB	
	Male	Female	Male	Female	Male	Female
P. flavus, Rodeo versus						
P. merriami, Uvalde	456	440	19.7	16.8	24.6**	13.0**
P. flavescens, Kearney	623	1012	13.2	12.1	12.8**	25.0**
P. flavus, Jornada versus						
P. merriami, Wink	700	570	28.9	28.6	32.6	28.3
P. merriami, Uvalde versus						
P. flavus, Rodeo	308	526	27.0	26.6	27.3	29.5
P. flavescens, Wink	767	653	6.9**	15.8**	4.3	6.7
P. flavescens, Kearney	427	1054	3.0**	8.2**	0.9**	28.4**
P. merriami, Wink versus						
P. flavus, Jornada	904	511	11.2**	22.7**	35.8**	16.6**
P. flavescens, Wink	843	1752	1.7**	12.3**	8.7	8.9
P. flavescens, Kearney	889	482	33.0	32.5	20.5	16.0
P. flavescens, Kearney versus						
P. flavus, Rodeo	997	442	4.5**	17.9**	37.9**	13.6**
P. merriami, Wink	808	910	5.6	8.5	48.0**	36.1**
P. merriami, Uvalde	777	431	1.4**	33.0**	42.6**	3.9**
P. flavescens, Wink versus						
P. merriami, Uvalde	1290	970	14.7	11.8	22.7	20.9
P. merriami, Wink	770	513	6.1**	11.8**	14.2**	26.9**

*P≤0.05.
**P≤0.01.

(Table 2), male monad frequencies were significantly greater than female frequencies in five encounters and significantly lower than female frequencies in two encounters. Male and female frequencies of WASH were not significantly different ($P>0.05$) in all encounters except female *P. flavescens*, Kearney, paired with male *P. merriami*, Uvalde, where female frequencies were significantly greater (10.4 versus 3.5, $P<0.01$).

Comparisons of sympatric male and female dyad frequencies in homospecific individuals revealed few differences (Martin, 1974). The dyad sandbathing followed by sandbathing (SNBH—SNBH), was directive, that is, occured significantly greater than expected ($P<0.01$), in most intrasubject (AA and BB) male and female comparisons with the exception of male *P. flavescens*, Kearney and male *P. flavescens*, Wink. The dyad DGKB—DGKB was also directive ($P<0.01$) in both AB and BA comparisons. The dyad NOCH—DBKB was directive ($P<0.01$) in all six AB comparisons and in three BA comparisons. The dyad NOCH—WASH was directive ($P<0.01$ and $P<0.05$) in two BA comparisons (female *P. merriami*, Uvalde, versus male *P. merriami*, Uvalde, and female *P. flavus*, Jornada, versus male *P. flavus*, Jornada).

Dyad frequencies of SNBH and DGKB with homospecific, allopatric individuals were similar to those from sympatric localities. Frequencies of

SNBH—SNBH were not significantly different $(P>0.05)$ from expected values in male *P. merriami*, Wink, and female *P. flavescens*, Wink. In female *P. merriami*, Uvalde, the frequency of DGKB—DGKB was not significantly different from the expected value. Intersubject comparisons of male-female dyad frequencies of SNBH and DGKB for allopatric homospecifics were similar to those of the sympatric homospecifics (for details see Martin, 1974).

Agonistic Behavioral Patterns

Weights, initial approaches and movements of Ss in homospecific encounters were compared for evidence that those variables might be indicators of dominant—subordinate rankings. The null hypotheses tested were that dominant individuals have equal probabilities of being larger or smaller than subordinates, making initial or later approaches, and making initial or later movements, compared with subordinates. When data for the similar-sized *P. flavus* and *P. merriami* were combined and compared with the larger species, *P. flavescens*, the null hypotheses for weight ($N=35$) and initial movements ($N=34$) were accepted ($P>0.25$, two-tailed binomial test, Siegel, 1956). But dominant individuals of *P. flavus* and *P. merriami* generally made the first approach in an encounter ($P<0.05$) although first approach was not a good indicator in *P. flavescens* $(P>0.40)$.

Monad frequencies of all variables were compared between subordinate and dominant individuals of the three species (Table 3). With strictly agonistic variables (Eisenberg,1967b), the frequencies of uprights (UPRG), chasing (CHAS), flight (FLGH), approach (APPR), and attack-escape leaps (ATES) were generally significantly different ($P<0.01$ and $P<0.05$) in dominant and subordinate individuals (Table 3). Significantly higher frequencies of FLGH and ATES (interpreted as escape leaps) were characteristic of male subordinates but not female subordinates (Table 3). Dominant individuals also had significantly greater frequencies of APPR in all combinations except those involving male *P. flavus* (Table 3). Significantly greater frequencies of CHAS occurred only in dominant females whereas significantly greater frequencies of UPRG occurred only in dominant males (Table 3).

The frequencies of SNBH, WASH, and marking (MARK) were not significantly different $(P>0.05)$ between dominant and subordinate males and females. But the frequencies of DGKB were significantly higher for dominant individuals and lower for subordinate individuals in all combinations except male *P. flavus* and female *P. merriami* (Table 3). No consistent differences in contact and sexual behavior were apparent between dominant and subordinate individuals. Dominant male *P. flavescens*, however, had significantly greater frequencies of circling (CRCL, 12.9 versus 4.2, $P<0.01$) and following and driving of females (FLDR, 6.1 versus 1.6, $P<0.05$) than did subordinate males.

Dyad frequencies for homospecific encounters (localities combined were compared for all possible intrasubject combinations (Figs. 2 to 4) and two

TABLE 3.—*Behavioral responses (monad frequencies) of dominant and subordinate individuals of* Perognathus *to homospecifics of a different sex. N is the total number of events.*

Behavioral Variable	P. flavus		P. merriami		P. flavescens	
	Dominant	Subordinate	Dominant	Subordinate	Dominant	Subordinate
Responses of Males						
N	287	358	687	162	309	1390
UPRG	13.4**	7.2*	12.0**	1.2**	34.8**	14.6**
CHAS	0.0	0.0	0.3	0.0	0.0	0.0
FLGH	1.7**	18.2**	3.1**	45.1**	1.6**	31.2**
APPR	11.3	6.6	12.0**	1.2**	8.4*	3.4*
ATES	0.3*	6.6*	0.3**	20.7**	0.7**	17.1**
DGKB	35.1	33.1	30.6**	4.3**	19.7**	11.6**
Responses of Females						
N	539	191	252	490	1867	198
UPRG	16.2	20.0	16.9**	7.1**	21.9	18.6
CHAS	4.8**	0.0**	11.4**	0.2**	6.5**	0.0**
FLGH	4.2**	13.9**	1.6**	13.5**	2.1**	16.6**
APPR	15.1**	2.1**	16.9**	7.1**	22.4*	13.1*
ATES	0.6	2.1	0.0	2.4	1.1	3.0
DGKB	19.2*	9.7*	37.8	28.6	14.7**	3.5**

*$P \leq 0.05$.
**$P \leq 0.01$.

intersubject combinations (*P. flavescens*, Wink, Fig. 5). In intrasubject comparisons, the dyads SNBH—SNBH, DGKB—DGKB, and UPRG—UPRG were generally directive ($P < 0.01$ or $P < 0.05$) for both dominant and subordinate inidvduals. But UPRG—UPRG was neither directive nor inhibitive in subordinate male *P. flavus* and *P. merriami*. Similarly, DGKB—DGKB frequencies were not significantly different from expected values in subordinate male *P. merriami* and subordinate *P. flavus* and *P. flavescens*.

The frequency of the dyad SDFG—SDFG was generally a good indicator of the dominant or subordinate status of individuals, with subordinate male *P. merriami* and female *P. flavescens* the exceptions. Repetitive SDFG was usually a trait of subordinates (Figs. 2 to 4).

In intrasubject comparisons, some dyads of the k—j type revealed different frequencies and conditional probabilities for dominant and subordinate Ss (Figs. 2 to 4). Most differences involved aggressive components such as APPR, CHAS, FLGH, and ATES; however, some of the dyads involved UPRG and DGKB. The frequencies of the dyads APPR—CHAS and CHAS—APPR were directive only in dominant female *P. flavescens* ($P < 0.01$). In subordinate Ss, frequencies were directive only for the dyads FLGH—APPR, in female *P. merriami* ($P < 0.01$), and ATES—FLGH, in male *P. merriami* ($P < 0.01$). Subordinate female *P. merriami* had an inhibitive frequency ($P < 0.01$) for the dyad APPR—FLGH.

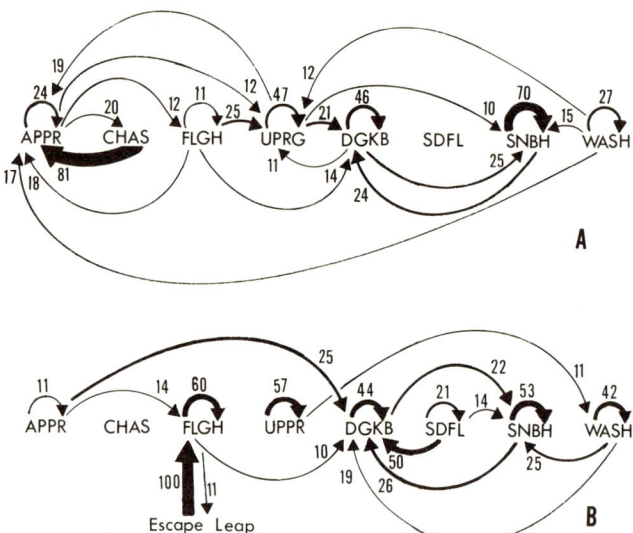

Fig. 2.—Flow diagrams, illustrating the most prominent behavioral variables exhibited by dominant (A) and subordinate (B) individuals of *Perognathus flavus* in intrasubject comparisons. The numbers in the figure refer to the transitional probabilities (converted to percentages) for particular dyads (that is, the occurrence of one behavioral variable followed by another variable). Lines were not drawn nor were percentages given when particular dyads occurred less than five times or the transitional probabilities for such dyads were less than 0.10. Line widths correspond to associated percentages.

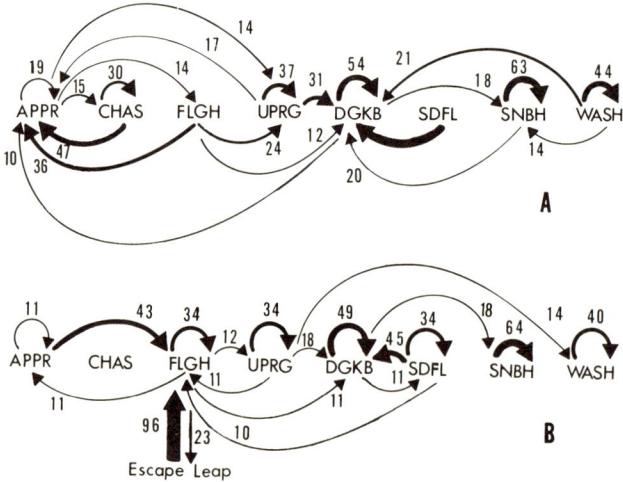

Fig. 3.—Flow diagrams, illustrating the most prominent behavioral variables exhibited by dominant (A) and subordinate (B) individuals of *Perognathus merriami* in intrasubject comparisons. See Fig. 2 for explanation of diagram.

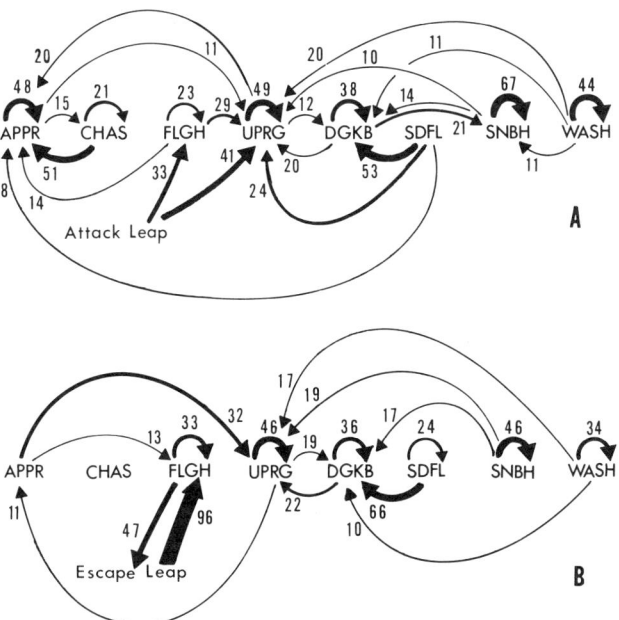

FIG. 4.—Flow diagrams, illustrating the most prominent behavioral variables exhibited by dominant (A) and subordinate (B) individuals of *Perognathus flavescens* in intrasubject comparisons. See Fig. 2 for explanation of diagram.

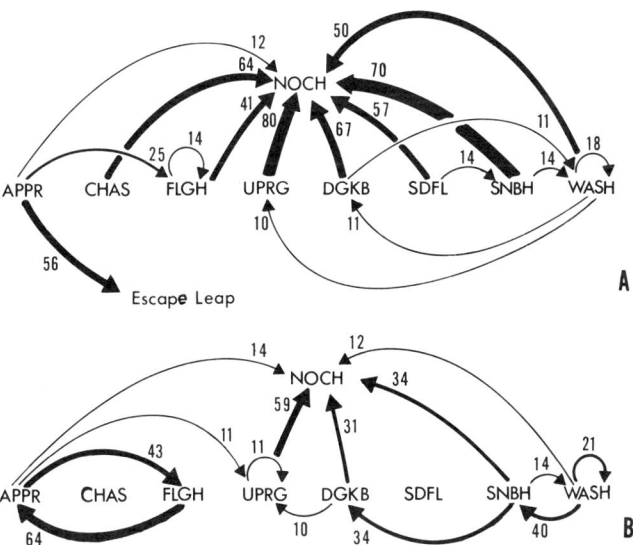

FIG. 5.—Flow diagrams, illustrating the most prominent behavioral variables exhibited by *Perognathus flavescens* in dominant-initiated (A) and subordinate-initiated (B) dyads of intersubject comparisons. See Fig. 2 for explanation of diagram.

In intersubject comparisons, dyads of the k—k type (that is, the display of a particular variable by one S followed by the display of the same variable by the other S) rarely occurred (Fig. 5). However, WASH—WASH frequencies were directive ($P<0.01$) although SNBH—SNBH frequencies were inhibitive ($P<0.05$). Dyads including no change (NOCH) occurred more frequently than any other dyads in intersubject comparisons. The actions of Ss usually produced no immediate change in the behavior of subordinate Ss (Fig. 5). In additon, the action of subordinate Ss upon the immediately following behavior of dominant Ss produced few directive dyads of the k—NOCH type (Fig. 5). In many instances, the actions of subordinate Ss produced responses other than NOCH in the dominant Ss (Fig. 5). Less freqently, the actions of dominant Ss were followed by variables other than NOCH (Fig. 5).

Frequencies of the dyads WASH—SNBH and FLGH—APPR were directive ($P<0.01$) in subordinate—dominant intersubject comparisons (Fig. 5B). The frequency of the dyad APPR—FLGH was directive ($P<0.01$) in subordinate—dominant comparisons (Fig. 5A). Examination of conditional probability values for these and one other dyad (SNBH—DGKB) also demonstrated differences between dominant- and subordinate-initiated behavior (Fig. 5). Thus, displays of SNBH, WASH, APPR, and FLGH by dominant Ss result in fewer displays by subordinates of DGKB, SNBH, FLGH, and APPR. Occurrences of these same patterns result in more displays of DGKB, SNBH, FLGH, and APPR by the dominant Ss in subordinate-initiated dyads.

Species and Locality Behavioral Patterns

Homospecific, heterospecific, and locality comparisons revealed strong similarities in the behavioral patterns of the smaller species, *Perognathus flavus* and *P. merriami*. But the behavioral patterns of these two species differed significantly from those of the larger species, *P. falvescens*. These differences were most evident when multivariate discriminant analyses were run on monad behavioral frequencies of the following fifteen variables: WASH, SNBH, APPR, NNAL, FLGH, ATES, UPRG, DGKB, HOUN, CRCL, CHAS, LCFG, SPAR, SDFG, GMPT. The multivariate analyses proved more useful than univariate inspection of data in summarizing complex interactions and simplifying interpretations. In the multivariate analyses, differences between groups were compared using D-statistics, F-values, and associated probability values. Standardized canonical coefficients for the 15 variables were listed in Martin (1974). Results of these analyses are presented below.

Allopatric

In allopatric tests, behavioral patterns were examined in populations of *P. flavus* from Rodeo, *P. merriami* from Uvalde, and *P. flavescens* from

Kearney (Fig. 1). Monad and dyad frequencies were similar, although some significant species differences were evident in the frequencies of DGKB and SNBH (Table 2). Allopatric *P. flavescens*, compared with *P. merriami* and *P. flavus*, had lower frequencies of DGKB, SNBH, and UPRG, and higher frequencies of the dyads DGKB—UPRG and SNBH—UPRG, in both hetereospecific and homospecific encounters.

In a multivariate three-group homospecific test, behavioral patterns of *P. flavus* and *P. merriami* overlapped considerably (average D=2.67, $F_{13,33}$=0.46, $P>0.05$), although each was clearly distinct from the patterns of *P. flavus* (*P. flavus* versus *P. flavescens*: D=6.81, $F_{13,33}$=14.57, $P<0.001$; *P. merriami* versus *P. flavescens*: D=7.08, $F_{13,33}$=15.61, $P<0.001$). *P. flavescens* was correctly classified into its own group in 15 of 16 cases (93.8 per cent), whereas *P. flavus* and *P. merriami* were frequently misclassified among themselves (5 of 16 *P. flavus* classified with *P. merriami* and 9 of 16 *P. merriami* classified with *P. flavus*). In this test, 97.8 per cent of the variation was accounted for by the first canonical variable (V_I). In V_I, ATES, HOUN, CRCL, UPRG, and WASH had high loadings (78.2 per cent of standardized coefficients), while V_{II} had high loadings on FLGH and HOUN (59.1 per cent).

In a multivariate four-group heterospecific test (Table 4), patterns of *P. flavus* paired with *P. merriami* (group 1) and *P. merriami* paired with *P. flavus* (group 2) were very similar. The patterns of groups one and two were distinctly separate from the patterns expressed in *P. flavescens* paired with *P. merriami* (group 3) and *P. merriami* paired with *P. flavescens* (group 4). The first two canonical variables accounted for 78.9 per cent and 16.0 per cent of the total variation, respectively. In V_I, NNAL, CRCL, CHAS, and LCFG had high loadings (52.4 per cent of standardized coefficients) while APPR, FLGH, CHAS, and LCFG had high loadings in V_{II} (75.1 per cent of standardized coefficients).

The high loadings on certain agonistic variables (CHAS, LCFG, APPR, and FLGH) reflect the importance of these variables in heterospecific encounters. Separation of *P. flavescens* from the two smaller species was also evident when dominance relationships were compared. The larger species was dominant in 23 of 32 heterospecific encounters. But dominance was about equally divided between *P. flavus* and *P. merriami* in heterospecific encounters involving these two species.

Near Contact and Jornada

Behavioral patterns were examined in mice collected near areas of contact that approached parapatry (*P. flavescens* and *P. merriami*, near Wink, Fig. 1) and from an area in New Mexico (Jornada, Fig. 1). Monad and dyad frequencies of behavioral patterns exhibited by the species from these areas were similar to the patterns expressed in the allopatric populations. In *P. flavescens*, in both heterospecific and homospecific encounters, monad

TABLE 4.—*Matrices of average distance (Mahalonobis D) of behavioral patterns, associated F-statistics (in parentheses), and per cent classified to group in a four-group heterospecific test. Group 1, Perognathus flavus paired with P. merriami; group 2, P. merriami paired with P. flavus; group 3, P. flavescens paired with P. merriami; group 4, P. merriami paired with P. flavescens. Allopatric populations.*

	Average D-values (F-statistics)		
Group	1	2	3
2	3.30 (0.95)		
3	4.66 (5.45)***	4.92 (7.39)***	
4	4.17 (3.10)**	3.90 (2.51)**	5.81 (11.11)***

	Classification Matrix with Per Cent Classified to Group				
Group	1	2	3	4	%
1	10	6	0	0	62.5
2	3	12	0	1	75.0
3	3	1	12	0	75.0
4	1	4	0	11	68.8

**$P \leq 0.01$

***$P \leq 0.001$; df=12, 49

frequencies of SNBH, DGKB, and UPRG were lower and dyad frequencies of SNBH—UPRG and DGKB—UPRG were higher. But the dyad frequencies of SNBH—UPRG in male *P. flavescens* were similar to those in the other two species.

In a multivariate three-group homospecific test, behavioral patterns of *P. flavus* and *P. merriami* were very similar (D=3.39, $F_{13,33}=0.69$, $P>0.05$) although each was significantly different from the pattern of *P. flavescens* (*P flavus* versus *P. flavescens*: D=4.98, $F_{13,33}=4.25$, $P<0.001$; *P. merriami* versus *P. flavescens*: D=4.88, $F_{13,33}=3.72$, $P<0.001$). *P. flavescens* was correctly classified in its own group in 13 of 16 cases (81.3 per cent), whereas *P. flavus* and *P. merriami* were frequently misclassified among themselves, although never with *P. flavescens* (4 of 16 *P. flavus* classified with *P. merriami* and 6 of 16 *P. merriami* classified with *P. flavus*). Compared with the allopatric populations at Uvalde and Rodeo, the Wink and Jornada populations of *P. merriami* and *P. flavus* were classified correctly (in a statistical sense) slightly more times (*P. merriami* and *P. flavus*: Allopatric, 43.8 and 68.8 per cent, respectively; near contact and Jornada, 62.5 and 75.0 per cent, respectively). But *P. flavescens* from Kearney were correctly classified more frequently (93.8 per cent) than those from the Wink populations (81.3 per cent). The first two canonical variables accounted for 88.4 and 11.6 per cent of all variation, respectively. In V_I, six, mostly agonistic, variables (APPR, NNAL, FLGH, UPRG, LCFG, SPAR) had the highest loadings (75.1 per cent of standardized coefficients). Loadings in V_{II} were high on two agonistic variables (LCFG, SPAR, 38.7 per cent) and three nonagonistic variables (WASH, SNBH, DGKB, 34.1 per cent).

TABLE 5.—*Matrices of average distance (D) of behavioral patterns, associated F-statistics (in parentheses), and per cent classified to group in a four-group heterospecific test. Group 1, P. flavus paired with P. merriami; group 2, P. merriami paired with P. flavus; group 3, P. flavescens paired with P. merriami; group 4, P. merriami paired with P. flavescens. Near contact and Jornada populations.*

Group	\multicolumn{3}{c}{Average D-values (F-statistics)}		
	1	2	3
2	3.14 (1.25)		
3	4.20 (3.06)**	4.38 (3.21)**	
4	3.77 (2.93)**	3.63 (1.76)	4.86 (4.82)***

Group	\multicolumn{5}{c}{Classification Matrix with Per Cent Classified to Group}				
	1	2	3	4	%
1	14	2	0	0	87.5
2	5	9	0	2	56.3
3	2	3	11	0	68.8
4	2	4	0	10	62.5

**$P \leq 0.01$.

***$P \leq 0.001$; df=12, 49.

In a mulitvariate four-group heterospecific test (Table 5), patterns of *P. flavus* paired with *P. merriami* (group 1) and *P. merriami* paired with *P. flavus* (group 2) were very similar. The patterns of groups one and two were generally distinctly separate from the patterns expressed in *P. flavescens* paired with *P. merriami* (group 3) and *P. merriami* paired with *P. flavescens* (group 4). The first three canonical variables accounted for 59.1, 29.4, and 11.5 per cent of the total variation, respectively, with agonistic variables having the highest loadings. In V_I, APPR, FLGH, and LCFG were most heavily weighted (76.0 per cent of standardized coefficients), followed by DGKB (12.1 per cent). The loadings of agonistic variables in V_{II} and V_{III} were less heavily weighted (59.1 per cent in V_{II}; 26.5 per cent in V_{III}). The variables DGKB and WASH had relatively high loadings in V_{III}(27.3 per cent).

An additional multivariate test (Fig. 6, Table 6) was made using six groups of two species (*P. merriami* and *P. flavescens*) to check for possible evidence of character displacement in behavioral patterns in areas of near contact that approximated parapatry (see discussion). The patterns of the following groups were examined:

Group 1.—Response of *P. merriami*, Uvalde, paired with *P. flavescens*, Kearney

Group 2.—Response of *P. flavescens*, Kearney, paired with *P. merriami*, Uvalde.

Group 3.—Response of *P. merriami*, Wink paired with *P. flavescens*, Kearney, and *P. merriami*, Uvalde, paired with *P. flavescens*, Wink

Group 4.—Response of *P. flavescens*, Wink, paired with *P. flavescens*, Kearney, and *P. flavescens*, Kearney, paired with *P. merriami*, Uvalde

Group 5.—Response of *P. merriami*, Wink paired with *P. flavescens*, Kearney

Group 6.—Response of *P. flavescens*, Wink, paired with *P. merriami*, Wink

TABLE 6.—*Matrices of average distance (D) of behavioral patterns, associated F-statistics (in parentheses), and per cent classified to group in a six-group heterospecific test. Group 1, P. merriami (Uvalde) paired with P. flavescens (Kearney); group 2, P. flavescens (Kearney) paired with P. merriami (Uvalde); group 3, P. merriami (Wink) paired with P. flavescens (Kearney) and P. merriami (Uvalde) paired with P. flavescens (Wink); group 4, P. flavescens (Wink) paired with P. flavescens (Kearney) and P. flavescens (Kearney) paired with P. merriami (Uvalde); group 5, P. merriami (Wink paired with P. flavescens (Kearney); group 6, P. flavescens (Wink) paired with P. merriami (Wink).*

Group	Average D-values (F-statistics)				
	1	2	3	4	5
2	3.95 (3.28)***				
3	3.57 (0.96)	4.42 (4.85)***			
4	4.51 (4.81)***	3.91 (1.30)	4.78 (6.27)***		
5	3.51 (1.26	4.30 (3.90)***	3.81 (1.02)	4.84 (5.63)***	
6	4.82 (4.73)***	4.81 (4.05)***	5.13 (5.43)***	5.53 (7.09)***	4.86 (3.66)***

Classification Matrix with Per Cent Correctly Classified to Group

Group	1	2	3	4	5	6	%
1	14	0	1	0	1	0	87.5
2	2	8	1	3	1	1	50.0
3	9	1	13	2	6	1	40.6
4	1	4	3	24	0	0	75.0
5	2	1	2	1	9	1	56.2
6	4	0	0	0	0	12	75.0

**$P \leq 0.01$.

***$P \leq 0.001$; df=15, 108.

Perognathus merriami from allopatric (group 1, Uvalde), near contact and allopatric (group 3, Wink and Uvalde) and near contact (group 5, Wink) exhibited no signficant differences in behavioral patterns when these populations were compared with one another. Similar comparisons in *P. flavescens* from allopatric (group 2, Kearney), near contact and allopatric (group 4, Wink and Kearney), and near contact (group 6, Wink) were significantly different from one another except for one comparison (group 2 versus 4). All other combinations involving heterospecific comparisons were signficantly different ($P < 0.001$).

Behavioral patterns were most similar within each species, yielding a cluster of three means for *P. merriami* and a cluster of three means for *P. flavescens* (Fig. 6). Comparisons of D-values between groups in these two clusters revealed that allopatric *P. merriami* (group 1) and *P. flavescens* (group 2) were most similar, behaviorally, (D=3.95) than were behavioral patterns between near contact populations of the two species (group 5 versus 6, D=4.86). Comparisons involving near contact and allopatric populations (group 3 versus 4) revealed behavioral patterns intermediate (D=4.78) between strictly allopatric (D=3.95) and near contact (D=4.86) populations.

In the six-group test, the first three canonical variables accounted for 53.3, 31.9, and 8.1 per cent of the total variation, respectively. Three nonagonistic

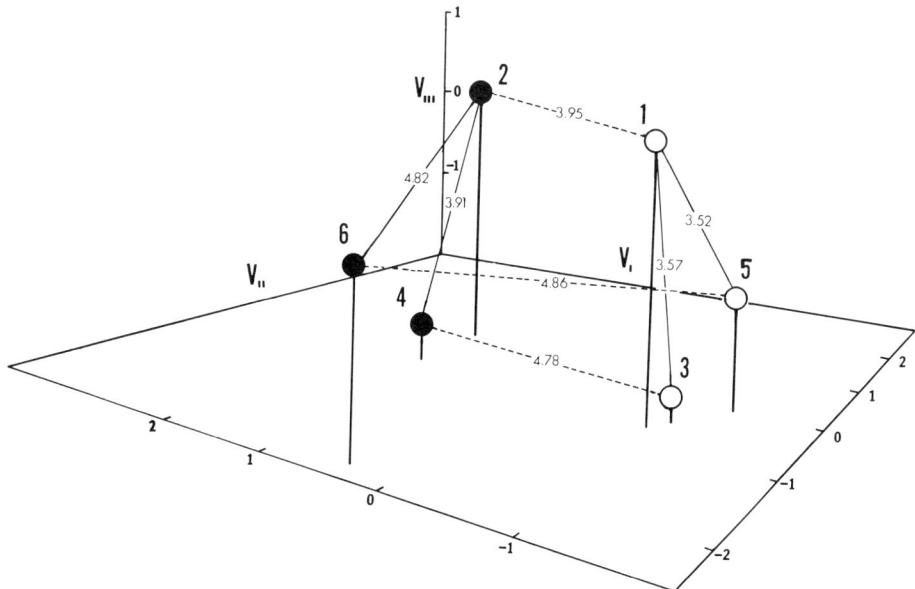

Fig. 6.—Projections of group means for behavioral responses of *Perognathus merriami* and *P. flavescens* onto the first three canonical variables (V_I, V_{II}, V_{III}). Solid lines link pairs of groups with the smallest D-values; dotted lines indicate other linkage relationships. Group 1, response of allopatric (Uvalde) *P. merriami*; 2, allopatric (Kearney) P. flavescens; 3, near contact (Wink) and allopatric (Kearney) *P. merriami*; 4, near contact (Wink) and allopatric (Kearney) *P. flavescens*; 5, near contact (Wink) *P. merriami*; 6, near contact *P. flavescens*.

variables (WASH, CRCL, GMPT) had the highest loadings (54.2 per cent of standardized coefficients) in V_I, whereas four variables (WASH, SNBH, UPRG, and GMPT) had high loadings in V_{II} (66.4 per cent). Three agonistic variables (LCFG, CHAS, SPAR) had relatively high loadings in V_{III} (24.8 per cent).

Comparison of dominance relationships in heterospecific trials of near contact and allopatric populations yielded results similar to those of strictly allopatric comparisons (Martin, 1974). But *P. flavus*, Jornada, were dominant in 10 of 16 trials with *P. merriami*, Wink. In similar tests with allopatric populations of *P. flavus* and *P. merriami* (Rodeo and Uvalde, respectively), each species showed dominance about equally (5 in *P. flavus*, 4 in *P. merriami*). In almost all cases (39 of 42), *P. flavescens* were dominant over *P. merriami* in both allopatric and near contact comparisons.

DISCUSSION

Comparison of behavioral patterns between different taxa or sexes frequently reveal few qualitative or absolute differences (Eisenberg, 1967b). Consequently, behavioral comparisons in species of mammals usually employ quantitative tabulations of varied complexity. Encounter experiments utilizing two subjects (for example, male versus female in this study) simplify the social situation and permit examination of many combinations of variables (Eisenberg, 1967b). These experimental conditions provide relative measures for comparisons and quantitative details that might be difficult to obtain under natural conditions.

Male and Female Behavioral Patterns

It is apparent from Table 2 and Martin (1974) that the null hypothesis of independence (equal probability) between the occurrences of each variable was rejected for both monad and dyad frequencies. Thus, a few variables accounted for a large portion of total behavioral events.

In 10 trials with *Perognathus inornatus*, Eisenberg (1963c) found males sandbathed more often than did females in paired encounters. Eisenberg's (1963a) results with *Dipodomys panamintinus* were similar, although the values he obtained with *D. merriami* revealed no apparent sexual differences in sandbathing frequencies. Eisenberg (1963a) suggested that the extreme agressive behavior exhibited by the male *D. panamintinus* could have been unusual. In the present study, inspection of SNBH frequencies of dominant and subordinate individuals suggested that relative aggression was not a major factor in distorting the frequency values of this variable. Rather, relative monad frequencies indicate a potential signaling mechanism for males to recognize females and vice versa.

Another function, perhaps related to visual communication, involves communication by chemical signals. Eisenberg (1963a, 1963c, 1967a, 1967b) stressed the importance of sandbathing as a potential means for depositing urine or sebum signals. Thiessen, *et al.* (1971) reported that male Mongolian gerbils (*Meriones unguiculatus*) marked twice as frequently as did females. Thus, if SNBH also functions as marking device, as Eisenberg suggested, the female *Perognathus* observed in this study leave more signals, or else intersperse chemical signaling with visual communication (that is, SNBH). The observation that the dyad SNBH—NOCH and the reciprocal NOCH—SNBH were directive suggests that the visual communication function of SNBH could be minimal; however, no change in one individual could serve as an appropriate signal to the other individual. Alternatively, a dyad analysis might be unable to detect more complex sequences that elicit appropriate behavior in another individual (for example, the quadrad a,f,c,g might result in a particular behavioral expression in the other individual). In studies of visual communication (displays) of hermit crabs, Hazlett and Bossert (1965) thought that sequence complexity was a more likely explanation of dyads of the k-no change type.

Agonistic Behavioral Patterns

Weights and initial movements of dominant individuals were poor predictors of "winning." Small individuals frequently dominated larger individuals. Likewise, males frequently inititated movements although they were not necessarily dominant in an encounter. Initial approach was generally associated with the eventual winner of an encounter.

Fleming (1974) found that larger individuals of *Liomys* were generally dominant in encounters, although this pattern did not hold for *Heteromys*. Similarly, Payne and Swanson (1970), found that larger hamsters were generally dominant. Grant (1970), found similar patterns in intraspecific tests with *Microtus*, *Peromyscus*, and *Clethrionomys*, as did Nevo *et al.* (1975) with the fossorial *Spalax* and Blaustein and Risser (1976) with three species of *Dipodomys*. The larger-animal-is-dominant hypothesis was not supported in the present study nor in studies involving two species of *Microtus* (Getz, 1962), *Microtus* and *Dicrostonyx* (Banks and Fox, 1968), and the lagomorph *Oryctolagus* (Mykytowycz and Hesterman, 1975). Thus, size might be an unreliable means for animals to ascertain the strength of an opponent.

Initial approach behavior was associated with eventual dominance in two species, *Perognathus flavus* and *P. merriami*. Female *P. flavescens* generally did not make the first approach although they were dominant in 14 of 16 trials. Fleming (1974) found that initial approaches generally predicted eventual outcomes of encounters, whereas female approaches had slightly less predictability in determining eventual dominance outcomes compared with male approaches. Perhaps female *P. flavescens* have higher thresholds for initiating approaches although they exhibited higher intensities of combativeness once the thresholds were reached. Clarke (1956) stated that nursing or late partum female voles were generally more aggressive than were males or nonpregnant females. Neither of these explanations are appropriate for explaining the extreme aggressiveness exhibited by female *P. flavescens*. Happold (1976) found that female conilurine rodents were generally the aggressors in most male-female encounters that she observed.

Dominant and subordinate individuals exhibited different behavioral patterns in encounters. Dominant individuals generally engaged in more approaches, chases, and uprights, whereas subordinates had higher frequencies of escape leaps and flight behavior. These patterns have been noted in many rodents, including other heteromyids (Eisenberg, 1963a; 1967a; Fleming, 1974; Ambrose and Meehan, 1977), microtines (Clarke, 1956; Murie and Dickinson, 1973), cricetines (Eisenberg, 1963b; Moore, 1965; Payne and Swanson, 1970), and murids (Happold, 1976). Less is known about sequences of behavior that characterize dominant and subordinate individuals.

Examination of dyad frequencies and conditional probabilities revealed consistent differences for dominants and subordinates. Five dyads (APPR—CHAS, CHAS—APPR, FLGH—APPR, FLGH—UPRG, and to some extent, UPRG—DGKB) occurred with greater frequency in dominants, whereas the dyad FLGH—ATES was most common in subordinates. The action of a dominant generally reduced aggressive responses by subordinates, whereas movements of a suborinate generally produced approaches or chase behavior by the dominant. Lerwill and Makings (1971) found that dominant and subordinate hamsters followed specific sequences. For example, a dominant hamster generally showed a threat posture followed by an offensive sideways posture, whereas a subordinate would evade the approach of a dominant and then exhibit a full submissive posture.

The dyad UPRG—DGKB generally occurred most frequently in dominants although there was some overlap with subordinates in conditional probability values. Lerwill and Makings (1971) found significant associations between "nose-offensive upright posture" and "offensive upright posture—threat" in dominant hamsters. Perhaps upright postures in close association with other postures provide signals of dominance status prior to agonistic displays (chase, flight, attack—escape leap). These signals may represent behavior of special importance in intraspecific aggression. Displays that signal dominant or subordinate status might reduce fighting and injury to animals.

If aggressive displays are present in *Perognathus*, one would expect to see more examples of appeasement behavior in subordinates; however, other than flight by the subordinate, few appeasement postures were observed in subordinates. In several trials involving male *P. merriami* and female *P. flavescens* from Wink, the subordinate male exhibited a defeat (DFRT) posture (Eisenberg, 1963a) after repeated chasing and attempted biting by the dominant female. DFRT appeared to temporarily diminish the intensity of the dominant's attack; however, when the subordinate moved or fled, the chasing and attacks resumed. Some of the intensity of aggression may be a result of the enclosed experimental situation. Eisenberg (1967a) noted that solitary species of the genera *Perognathus*, *Liomys*, *Dipodomys*, and *Gerbillus* react with aggressive behavior or avoidance during encounters, particularly in male-nonestrus female encounters. Agonistic behavior (both aggressive and submissive) may function to limit contacts between individuals and minimize competition. MacMillen (1964) and Maza, *et al.* (1973) noted a general lack of overlap in individual home ranges of heteromyids, indicative of individual spacing. Maza, *et al.* further suggested that changes in the size of heteromyid home ranges were more a reflection of the frequency of social interaction rather than the quality of the environment. But aggressive individuals could conceivably "defend" or "secure" a larger territory and prevent its exploitation by another individual.

Simple avoidance behavior might be a factor for some of the high loadings on the variable FLGH. Visual observations demonstrated that two individuals would often make contact or near contact and then initiate aggressive behavior such as SDFL, CHAS, SPAR, LCFG, and ATES. Thus, simple avoidance behavior does not adequately explain the extreme aggression exhibited by these mice.

Species and Locality Behavioral Patterns

Behavioral patterns of all three species of *Perognathus* were basically similar. Similar trends were noted in studies involving two species of *Peromyscus* (Moore, 1965) and *Microtus* (Miller, 1969). It is known that species differences in behavioral patterns often operate at a quantitative rather than qualitative level (Mayr, 1963; Alexander, 1967; Johnsgard, 1972; Bekoff, *et al.*, 1975). Inspection of monad and dyad frequencies revealed distinctive quantitative patterns between the larger species (*Perognthus flavescens*) and the two smaller species (*P. flavus* and *P. merriami*). In both homospecific and heterospecific encounters, *P. flavescens* had greater frequencies of the dyads DGKB—UPRG and SNBH—UPRG compared with values for the smaller species. The patterns of these two dyads in homospecific encounters did not differ greatly between dominant or subordinate individuals; however, these differences serve as recognition devices between species or as indicators of dominant or subordinate status of potential combatants prior to aggressive displays. Under natural conditions, prefighting displays might reduce aggression and potential injury to both animals.

Multivariate analyses also revealed similarities in behavioral patterns of *P. flavus* and *P. merriami*; the larger species exhibited patterns distinct from the smaller species in both homospecific and heterospecific encounters. *Perognathus flavescens* was clearly dominant in nearly all heterospecific encounters. Large species frequently are dominant over smaller species (Moore, 1965; Grant, 1969; Sheppard, 1971) but not always (Getz, 1962; Banks and Fox, 1968; Grant, 1969).

Several differences were noted in comparisons of homospecific and heterospecific responses of allopatric and near contact populations. In allopatric comparisons, *Perognathus flavus* and *P. merriami* were classified to their respective groups less frequently than in comparisons involving near contact (*P. merriami* from Wink and *P. flavus* from Jornada). The differences in behavioral patterns observed did not seem sufficient to suspect any premating isolating mechanisms or specific distinctness for these two forms. These behavioral data substantiate the conclusions of Wilson (1973), Martin (1977), and Schmidly (1977) that these two species are conspecific.

Behavioral patterns involving *P. merriami* and *P. flavescens* revealed greater differences when the populations were from near contact localities (Wink) compared with their respective allopatric localities (Uvalde and Kearney). Several explanations for those results are plausible.

Brown and Wilson (1956) noted that morphological differences between related species can be greater in areas of sympatry than in those of allopatry. This character displacement or divergence could result from selection that 1) reinforces premating isolating mechanisms between the species, or 2) favors ecological divergence (Brown and Wilson, 1956; Mayr, 1963). Miller (1967), Bulmer (1974), Huey and Pianka (1974), and Slatkin (1980) noted that competition might promote the development of character displacement. And Nevo *et al.* (1975) reported that reinforcement of aggressive behavioral patterns was the most likely explanation for the parapatric distributions observed in *Spalax.*

The behavioral patterns of *P. merriami* and *P. flavescens* might represent an example of competitive exclusion. This mechanism is suggested although some data are presently lacking. To date, no quantitative estimates are available on the realized niches of these two species nor on the fundamental niches of each species in the experimental absence of the other. Also, parapatry was not demonstrated in the present study; the nearest adjacent sampling localities were 12 km. apart although R. L. Packard (personal communication) collected *P. merriami* within several hundred meters of where I captured *P. flavescens.*

In areas of contact, I believe that *Perognathus flavescens* represents the specialized species that has an included niche "within" the broader niche of the generalist, *P. merriami.* The specialist, in interactions with *P. merriami,* would be expected to 1) be more aggressive, 2) have a narrower tolerance for various environmental variables, and 3) exclude *P. merriami* from areas of deep sand.

Increased aggression was demonstrated in near contact interactions. In staged encounters, Sheppard (1971) and Heller (1971) found that *Eutamias amoenus,* with an included niche, was more aggressive than was *E. minimus,* which had the broader niche. Brown (1971) also noted similar patterns in *E. dorsalis* (the more aggressive species) and *E. umbrinus.* Kritzman (1974) demonstrated that *Peromyscus maniculatus* narrowed its diet when in contact with *Perognathus parvus.* Ambrose and Meehan (1977) suggested that this shift was due to the greater aggressiveness of *Perognathus parvus.*

Perognathus flavescens is near the southern end of its geographic range at Wink (Fig. 1), where it occupies a habitat of deep sand (Williams, 1971). Cameron (1971) noted that the specialist and more aggressive species, *Neotoma fuscipes,* occupied the most favorable habitat. At Wink, *P. merriami* was found in hardpan soils with a high clay content (Martin, 1974); it is rarely collected in deep sand (R. L. Packard, personal communication). Sheppard (1971) found that *Eutamias amoenus* and *E. minimus* exhibited greater habitat diversity in the absence of competing species. No comparable trends were noted within localities where *Perognathus* were collected; however, *P. merriami* and its probable conspecific, *P. flavus,* occur in many soil types throughout their ranges (Martin, 1974).

Relative population densities of the agressive *P. flavescens* were lower (2.2 mice per 100 trap nights) in the Wink area compared with allopatric populations (Kearney: 15.6 to 27.7 per 100 trap nights; Halsey: 16.6 per 100 traps nights). At Wink, relative population densities of *P. merriami* were higher (5.0 to 5.5 per 100 trap nights) than those of *P. flavescens.* Brown

(1971) noted that the more aggressive *E. dorsalis* had lower population densitities when in contact with *E. umbrinus.*

Evidence of exclusion between two species is not necessarily a demonstration of competitive exclusion (Mayr, 1963; Grant, 1972; Colwell and Fuentes, 1975; Slade and Robertson, 1977; Slatkin, 1980). Prior to secondary contact, these adaptations to particular habitats minimize interactions between species. Interactions that occur between species reflect incomplete adaptation to the particular environment, competition, or both. But Bowers and Brown (1982) suggested that contemporary ecological interactions among coexisting species rather than an historical legacy of geographic speciation was a more likely explanation for the patterns of negative association between pairs of similar-sized heteromyids. Also, Blaustein and Risser's (1976) work indicated that interspecific aggression is one mechanism by which ecological separation is maintained in three species of *Dipodomys* they studied.

It is not known whether the behavior observed at Wink represents character displacement, competitive exclusion, differences acquired prior to secondary contact, or combinations of these phenomena. I believe that selection favored increased aggressiveness of *P. flavescens* in the area of near contact. This species probably has a narrower niche than does *P. merriami* and thus excludes the latter species by aggressive behavior. Whether *P. flavescens* has a niche narrower than that of *P. merriami* must await additional studies. In particular, those studies should examine resource partitioning in allopatric and near contact populations. Experimental manipulations of field populations will also help to determine the role of competition in structuring communities of these small heteromyids.

SUMMARY

A few variables accounted for a large proportion of total behavioral events. Comparison of stochastic events and multivariate discriminate analyses proved useful in clarifying complex interactions of behavioral variables. These analyses are recommended to detect subtle differences of evolutionary significance since qualitative behavioral differences may not be sufficient at the species level.

In some species interactions, male and female behavior was similar although often quantitatively different in a few variables (sandbathing, digging and kicking back). These differences suggest that certain behaviors may function as signals between sexes and individuals.

Weights and initial movements were poor predictors of animals winning encounters although initial approaches by individuals were good indicators of dominance. Consistent and striking behavioral differences were noted for dominants and subordinates in both monad and dyad frequencies. The actions of dominants greatly reduced activity in subordinates. Conversely, increased aggressive displays were noted in dominants as a result of activity by subordinates.

Divergence in behavioral patterns was apparent in *Perognathus merriami* and *P. flavescens* from a near contact area. In contrast, patterns were most similar between these two species when individuals from distant allopatric populations were examined. Selection favoring extreme aggressive behavior in *P. flavescens*, suggested as the most specialized species, seems the most probable explanation for this divergence.

Behavior patterns of *P. flavus* and *P. merriami* were similar in monad, dyad, and multivariate analyses. Patterns of *P. flavescens* were significantly different from the patterns of the smaller species. These behaviors and the results of preference experiments (Martin, 1977) support Wilson's (1973) and Schmidly's (1977) conclusions that *P. flavus* and *P. merriami* are conspecific.

Acknowledgments

This paper is dedicated to Robert L. Packard who appraised the research design, provided laboratory space, and allowed me the freedom to pursue the directions that this study led. Funding was partially supported by a Grant-in-Aid of Research from the Society of the Sigma Xi, two summer fellowships from the Department of Biological Sciences, Texas Tech University, and a research assistantship for one year under the provisions of the IBP Grasslands Biomes studies supported by NSF (GB-31862X and GB-31862X2). An earlier version of this paper, which represents part of my dissertation for a Ph.D. degree from Texas Tech University, was read and evaluated by Archie C. Allen, John P. Brand, John S. Mecham, and Russell W. Strandtmann. Kenneth G. Matocha, Brian R. Chapman, Tony Mollhagen, Jeffrey Schultz, Robert W. Wiley, and Dan Womochel provided help and suggestions during the project. Duane Anderson advised on statistical analyses. My wife, Patty, aided in transcribing data. The help of all of these individuals and a reviewer is gratefully acknowledged.

Literature Cited

Alexander, R. D. 1967. Comparative animal behavior and systematics. Pp. 494-517, *in* Systematic biology (C. G. Sibley, chairman). Publication 1692, National Academy Sciences, Washington, 632 pp.

Altman, S. A. 1965. Sociobiology of Rhesus monkeys. II. Stochastics of social communication. J. Theor. Biol., 8:490-522.

Ambrose, R. F., and T. E. Meehan. 1977. Aggressive behavior of *Perognathus parvus* and *Peromyscus maniculatus*. J. Mamm., 58:665-668.

Aspey, W. P. 1977. Wolf spider sociobiology: I. Agonistic display and dominance-subordinance relations in adult male *Schizocosa crassipes*. Behaviour, 62:103-141.

Banks, E. M., and S. F. Fox. 1968. Relative aggression of two sympatric rodents: a preliminary report. Commun. Behav. Biol (Part A), 2:51-58.

Bekoff, M., H. L. Hill, and J. B. Mitton. 1975. Behavioral taxonomy in canids by discriminant function analyses. Science, 190:1223-1225.

Blaustein, A. R., and A. C. Risser, Jr. 1976. Interspecific interactions between three sympatric species of kangaroo rats (*Dipodomys*). Anim. Behav., 24:381-385.

BOWERS, M. A., AND J. A. BROWN. 1982. Body size and coexistence in desert rodents: chance or community structure. Ecology, 63:391-400.

BROWN, J. H. 1971. Mechanisms of competitive exclusion between two species of chipmunks. Ecology, 52:305-311.

BROWN, W. L., JR., AND E. O. WILSON. 1956. Character displacement. Syst. Zool., 5:49-64

BULMER, M. G. 1974. Density-dependent selection and character displacement. Amer. Nat., 108:45-58.

CAMERON, G. N. 1971. Niche overlap and competition in woodrats. J. Mamm., 52:288-296.

CHAPMAN, B. R., AND R. L. PACKARD. 1974. An ecological study of Merriam's pocket mouse in southeastern Texas. Southwestern Nat., 19:281-291.

CLARKE, J. R. 1956. The aggressive behaviour of the vole. Behaviour, 9:1-23.

COLWELL, R. K., AND E. R. FUENTES. 1975. Experimental studies of the niche. Ann. Rev. Ecol. Syst., 6:281-310.

DAVIS, W. B. 1974. The mammals of Texas. Bull. Texas Parks and Wildlife Department, 41(revised):1-294 pp.

DIXON, W. J. (ed.). 1971. BMD biomedical computer programs. Univ. California Publ. Automatic Computation, 2:v-x+1-600.

EISENBERG, J. F. 1963a. The behavior of heteromyid rodents. Univ. California Publ. Zool., 69:1-100+13 pls.

———. 1963b. The intraspecific social behavior of some cricetine rodents of the genus Peromyscus. Amer. Midland Nat., 69:240-246.

———. 1963c. A comparative study of sandbathing behavior in heteromyid rodents. Behavior, 22:16-23.

———. 1967a. A comparative study in rodent ethology with emphasis on evolution of social behavior, I. Proc. U.S. Nat. Mus., 122:1-51.

———. 1967b. The social organizations of mammals. Handbuch Zool. (vol. 8, Mammalia), 10(7):1-92.

FLEMING, T. H. 1974. Social organization in two species of Costa Rican heteromyid rodents. J. Mamm., 55:543-561.

FORBES, R. B. 1962. Notes on food of silky pocket mice. J. Mamm., 43:278-279.

FREY, D. F., AND R. J. MILLER. 1972. The establishment of dominance relationships in the blue gourami, *Trichogaster trichopterus* (Pallus). Behaviour, 42:8-62.

GETZ, L. L. .1962. Aggressive behavior of the meadow and prairie voles. J. Mamm., 43:351-358.

GRANT, P. R. 1969. Experimental studies of competitive interaction in a two-species system. I. *Microtus* and *Clethrionomys* species in enclosures. Canadian J. Zool., 47:1059-1082.

———. 1970. Experimental studies of competitive interaction in a two-species system. II. The behaviour of *Microtus*, *Peromyscus* and *Clethrionomys* species. Anim. Behav., 18:411-426.

———. 1972. Interspecific competition among rodents. Ann. Rev. Ecol. Syst., 3:79-106.

HALL, E. R. 1981. The mammals of North America. 2nd. ed. John Wiley and Sons, New York, 1:xviii+1-600+*90*.

HALL, E. R., AND K. R. KELSON. 1959. The mammals of North America. Ronald Press Co., New York, 1:xxx+1-546 +79.

HAPPOLD, M. 1976. Social behavior of the conilurine rodents (Muridae) of Australia. Z. Tierpsychol., 40:113-182.

HAZLETT, B. A., AND W. H. BOSSERT. 1965. A statistical analysis of the aggressive communications systems of some hermit crabs. Anim. Behav., 13:357-373.

HELLER, H. C. 1971. Altitudinal zonation of chipmunks (*Eutamias*): interspecific aggression. Ecology, 52:312-319.

HUEY, R. B., AND E. R. PIANKA. 1974. Ecological character displacement in a lizard. Amer. Zoo., 14:1127-1136.

JOHNSGARD, P. A. 1972. Animal behavior, 2nd. ed. Wm. C. Brown, Dubuque, viii+168 pp.

Kritzman, E. B. 1974. Ecological relationships of *Peromyscus maniculatus* and *Perognathus parvus* in eastern Washington. J. Mamm., 55:172-188.

Lerwill, C. J., and P. Makings. 1971. The agonistic behaviour of the golden hamster *Mesocricetus auratus* (Waterhouse). Anim. Behav., 19:714-721.

MacMillen, R. E. 1964. Population ecology, water relations, and social behavior of a southern California semidesert rodent fauna. Univ. California Publ. Zool., 71:1-66.

Martin, R. E. 1974. Behavioral patterns and species preferences in silky pocket mice, genus *Perognathus* (Rodentia: Heteromyidae). Unpublished Ph.D. dissertation, Texas Tech University, 184 pp.

——. 1977. Species preferences of allopatric and sympatric populations of silky pocket mice, genus Perognathus (Rodentia: Heteromyidae). Amer. Midland Nat., 98:124-136.

Mayr, E. 1963. Animal species and evolution. Harvard Univ. Press, Cambridge, xvi+797 pp.

Maza, B. G., N. R. French, and A. P. Aschwanden. 1973. Home range dynamics in a population of heteromyid rodents. J. Mamm., 54:405-425.

Miller, R. S. 1967. Pattern and process in competition. Adv. Ecol. Res., 4:1-74.

Miller, W. C. 1969. Ecological and ethological isolating mechanisms between Microtus pennsylvanicus and Microtus ochrogaster at Terre Haute, Indiana. Amer. Midland Nat., 82:140-148.

Moore, R. E. 1965. Ethological isolation between Peromyscus maniculatus and Peromyscus polionotus. Amer. Midland Nat., 74:341-349.

Murie, J. O., and D. Dickinson. 1973. Behavioral interactions between two species of red-backed vole (*Clethrionomys*) in captivity. Canadian Field-Nat., 87:123-129.

Mykytowycz, R., and E. R. Hesterman. 1975. An experimental study of aggression in captive European rabbits, *Oryctolagus cuniculus* (L.). Behaviour, 52:104-123, pls. xiii-xvii.

Myrberg, A. A., Jr. 1972. Ethology of the bicolor damselfish, *Eupomacentrus partitus* (Pisces: Pomacentridae): a comparative analysis of laboratory and field behaviour. Anim. Behav. Monog., 5:197-283.

Nelson, K. 1964. The temporal patterning of courtship behaviour in the glandulocaudine fishes (Ostariophysi, Characidae). Behaviour, 24: 90-146.

Nevo, E., G. Naftali, and R. Guttman. 1975. Aggression patterns and speciation. Proc. Nat. Acad. Sci., 72:3250-3254.

Osgood, W. H. 1900. Revision of the pocket mice of the genus Perognathus. N. Amer. Fauna, 18:1-73.

Patton, J. L. 1967. Chromosomes and evolutionary trends in the pocket mouse subgenus *Perognathus* (Rodentia: Heteromyidae). Southwestern Nat., 12:429-438.

Payne, A. P., and H. H. Swanson. 1970. Agonistic behaviour between pairs of hamsters of the same and opposite sex in a neutral observation area. Behaviour, 36:259-269.

Pimentel, R. A., and D. F. Frey. 1978. Multivariate analysis of variance and discriminant analysis. Pp. 247-274, in Quantitative ethology (P. W. Colgan, ed.). John Wiley and Sons, New York, xiv+364 pp.

Schmidly, D. J. 1977. The mammals of Trans-Pecos Texas. Texas A&M University Press, College Station, xiii+225 pp.

Siegel, S. 1956. Nonparametric statistics for the behavioral sciences. McGraw-Hill Book Co., New York, xvii+312 pp.

Sheppard, D. H. 1971. Competition between two chipmunk species (*Eutamias*). Ecology, 52:320-329.

Slade, N. A., and P. B. Robertson. 1977. Comments on competitively-induced disjunct allopatry. Occas. Pap. Mus. Nat. Hist. Univ. Kansas, 65:1-8.

Slatkin, M. 1980. Ecological character displacement. Ecology, 61:163-177.

Thiessen, D. D., G. Lindzey, S. L. Blum, and P. Wallace. 1971. Social interactions and scent marking in the Mongolian gerbil (*Meriones unguiculatus*). Anim. Behav., 19:505-513.

WILLIAMS, D. F   1971.   The systematics and evolution of the *Perognathus fasciatus* group of pocket mice.   Unpublished Ph.D. dissertation, Univ. New Mexico, 174 pp.

————.   1978.   Karyological affinities of the species groups of silky pocket mice (Rodentia, Heteromyidae). J. Mamm., 59:599-612.

WILSON, D. E.   1973.   The systematic status of *Perognathus merriami* Allen.   Proc. Biol. Soc. Washington, 86:175-192.

VARIATION OF THE TRILL CALLS OF
THREE SPECIES OF SPERMOPHILUS

KENNETH G. MATOCHA

Ground squirrels are well suited for behavioral observations as they are vociferous, diurnal rodents that generally inhabit relatively open areas. Several recent studies have focused on the vocalizations of *Spermophilus* and the behavior associated with those calls (Betts, 1976; Owings, *et al.*, 1977; Owings and Leger, 1980; Robinson, 1981; Sherman, 1977; Shields, 1980; and Turner, 1973). In spite of the recent work, most studies have not dealt with the variation of vocalizations in this genus. To determine mechanisms (for example, genetic or learned) by which vocal variation arose or the importance (for example, individual recognition or information content) of this variation, one must know the extent and nature of the variation present. The objectives of this study were to 1) describe and compare the trill calls of *Spermophilus tridecemlineatus*, *S. spilosoma annectens*, *S. s. marginatus*, and *S. mexicanus* and 2) evaluate the geographic variation in the trill call of *S. tridecemlineatus arenicola*.

MATERIALS AND METHODS

Ground squirrel vocalizations were recorded in their natural habitat with a Concertone 700 portable recorder on Soundcraft 1-mil magnetic tape at 19.05 centimeters per second (sec.). The recorder was fitted with a Geloso-M1/121 cardioid-unidirectional microphone. A Uher 6000L recorder was used to feed impulses into a Tektronix 564B storage oscilloscope fitted with a 3A75 amplifier and 2B67 time-base unit.

A stepwise discriminant analysis was performed on an IBM 370-145 digital computer using the BMD07M program of Dixon (1971). The following variables were used for the discriminant analysis:

Number of pulses (NP).—Total number of energy bursts in a single vocalization.

Pulse repetition rate (PRR).—Number of pulses minus one, divided by the length (in seconds) from the middle of the first to the middle of the last pulse.

Rise time (RT).—Length in milliseconds (msec.) from 10 per cent displacement for a single pulse.

Fall time (FT).—Length in msec. from peak amplitude to 10 per cent displacement for a single pulse.

Maximum energy pulse (MEP).—Pulse number from beginning of call and containing peak amplitude or maximum energy displacement for a given call.

Frequency (F).—A direct measurement of the dominant (strongest) frequency in cycles per sec. (Hertz, Hz.)

In addition, the following variables were also measured: call length (CL), pulse length (PL), interpulse length (IPL), call shape (CS), fall/rise ratio

(FFR) and pulse-interpulse ratio (PIR). A full description of these variables was given by Matocha (1977).

Descriptions and interspecific comparisons were made of trill vocalizations of *Spermophilus tridecemlineatus*, *S. spilosoma annectens*, *S. s. marginatus* and *S. mexicanus*. Calls of *S. tridecemlineatus* were from the Lubbock population (identified below). This population was chosen for these comparisons as it was geographically the closest to the other species studied. *S. mexicanus* calls were from individuals in Ward County, 0.1 mi. N *Monahans*, Texas. Calls of *S. spilosoma* were: *S. s. annectens*, Nueces County, 5 mi. S Port Aransas, Texas on Mustang *Island* and *S. s. marginatus*, Winkler County, 7 mi. WSW *Kermit*, Texas (Hall, 1981; Howell, 1938).

Intraspecific comparisons in 12 populations of *S. tridecemlineatus* were used to determine geographic variation in trill calls of adults from Texas, New Mexico, Colorado, Kansas, and Oklahoma as follows: Texas, 2 mi. S Woodrow, *Lubbock* Co.; New Mexico, Curry Co., city limits of *Clovis*, Union Co., 1 mi. E *Clayton*; Colorado, Otero Co., 3 mi. N *La Junta*, Washington Co., 1 mi. S *Akron*; Kansas, Cheyenne Co., city limits of *St. Francis*, Norton Co., 1 mi. S *Norton*, Scott Co., city limits of *Scott City*, Barton Co., 0.1 mi. N *Great Bend*, Barber Co., 1.5 mi. E *Medicine Lodge*, Oklahoma, Texas Co., city limits of *Guymon*, Woodward Co., city limits of *Woodward*. The names italicized above are used throughout the text to designate the populations at each locality.

A maximum of 20 calls recorded in the field from each population was selected for study. No attempt was made to determine the sex of the calling animal, but rather the calls were recorded at random. If these squirrels exhibit calling behaviors similar to *S. beldingi*, the samples may contain more vocalizations from females than from males (Sherman, 1977, 1980).

Variables used in the canonical analysis were Number of Pulses (NP), Pulse Repetition Rate (PRR), Rise Time (RT), Fall Time (FT), Frequency (F), and Maximum Energy Pulse (MEP). Some variables are present only when a call consists of two or more pulses (PRR and IPL). Since trill vocalizations of *S. s. annectens* often consisted of a single pulse, only NP, RT, FT, F and MEP were used in the canonical analysis for species comparisons. Sound spectograms of trill calls were made for visual comparisons.

RESULTS

Description of Trill Calls

Spermophilus tridecemlineatus had a mean call length (±SD) of slightly over one second (1.16±0.28 sec.), with a mean of 16.2 (±3.5) pulses per call (Table 1). The mean length of the pulse was 41.8 (±4.1) msec., with a rise time of 14.9 (±7.2) msec. Mean frequency of the trill call was 4580 (±112) Hz. Maximum energy was present in the first portion of the call and

TABLE 1.—*Group means and standard deviations of variables measured on trill vocalizations of* Spermophilus tridecemlineatus, S. spilosoma annectens, S. s. marginatus, *and* S. mexicanus.

Group	Number of pulses	Pulse repetition rate	Rise time	Fall time	Maximum energy pulse	Frequency
S. tridecemlineatus						
Akron	13.9±2.0	14.1±1.3	8.4±4.4	37.2±8.3	5.1±3.4	4983±263
La Junta	11.3±3.5	15.7±2.3	4.6±2.7	34.9±6.2	3.1±1.6	4668±93
St. Francis	19.7±4.8	13.6±1.2	8.9±5.9	37.9±9.2	3.2±1.8	4872±390
Norton	18.5±2.6	13.6±1.1	10.1±6.9	36.9±8.2	3.2±2.6	5094±334
Scott City	16.1±4.7	13.6±0.9	6.0±4.2	37.3±10.0	4.0±3.9	4993±316
Great Bend	16.7±2.6	14.7±0.8	3.7±2.2	38.6±7.1	3.6±2.3	4769±162
Medicine Lodge	14.2±2.1	11.7±0.9	11.3±7.1	38.6±10.6	2.3±1.4	4664±271
Guymon	11.1±2.1	12.3±1.4	5.4±3.3	37.9±8.6	4.4±3.8	4952±238
Woodward	17.3±3.3	11.9±1.0	13.2±4.9	30.8±7.9	2.1±0.9	4510±124
Clayton	11.7±1.6	11.9±1.3	9.2±7.0	40.3±9.1	4.2±2.0	4752±145
Clovis	14.3±2.3	14.6±1.0	7.8±5.8	36.3±5.7	2.9±1.8	4478±172
Lubbock	16.2±3.5	16.2±1.2	14.9±7.2	26.9±6.6	4.8±3.6	4581±112
S. s. annectens	3.3±2.2		12.0±11.9	36.4±17.2	1.7±0.8	5214±541
S. s. marginatus	28.4±2.1		6.9±3.9	25.4±4.3	5.3±4.8	5093±88
S. mexicanus	47.0±7.8		12.3±3.0	10.2±3.6	8.4±5.6	4002±215

gradually decreased until the end of the call (Fig. 1a). A sonagram of this call (Fig. 1a) shows a general decrease in frequency toward the end of the call.

The trill call (Fig. 1b) of *S. spilosoma annectens* (Island) was short (mean length of 0.26±0.18 sec.) with few pulses (3.3±2.2 per call) and, in about 35 per cent of the cases, consisted of only a single pulse. Of the four populations studied, Island *S. s. annectens* had the longest pulse length (49.3±17.6 msec.). Rise time was 12.0 msec. (±11.9). Frequency of trill calls of this population was the highest recorded (5214±541 Hz.). Interpulse length was longer than pulse length (54.2±7.2 msec.). Maximum amount of energy was usually contained in the first pulse of the call. The frequency remained fairly constant throughout the call. Fig. 1b shows a rapid increase in frequency followed by a decrease of frequency within each of the pulses produced within a given call.

The calls of individuals of *S. s. marginatus* from Kermit were intermediate between those of *S. tridecemlineatus* and *S. mexicanus* in length (1.6±0.13 sec.) and in number of pulses (28.4±2.1). Mean pulse length was 32.2±3.8 msec., and the rise time was short (6.9±3.9 msec.). Frequency was higher than that of individuals of both *S. tridecemlineatus* and *S. mexicanus* (5093±88 Hz.) and remained constant throughout the vocalization. Two variations of the trill call were noted in terms of the energy distribution, with a number of examples being intermediate. Oscillograms of the two variations are shown in Fig. 1c, d. There also were some differences noted in the sonagrams of these two variations of the trill call (see Fig. 1). The differences did not separate this population into two groups, as many of the calls were intermediate between the two extremes.

Individuals of *Spermophilus mexicanus* had the longest calls (2.6±0.5 sec.) of the four populations; a full second longer than those of *S. s. marginatus* from Kermit. Calls of *S. mexicanus* contained a mean of 47.0±7.7 pulses. Pulse length was short (22.5±2.6 msec.), with a long rise time (12.3±3.9 msec.). Vocalizations of *S. mexicanus* had the lowest frequency of any squirrel studied (4002±215 Hz.). Abrupt changes in energy were noted within calls of this species (Fig. 1e). The pulse containing maximum energy was located at the beginning of the call (average MEP = 9), and little energy was contained near the end. Frequency was lowest in the first pulse, quicky reaching a maximum frequency, then leveling off for the remainder of the call.

Interspecific Variation of Trill Calls

Discriminant analysis revealed only two dimensions (canonical variates) necessary for complete separation of the four populations (Lubbock, Island, Kermit, and Monahans). These two canonical variates accounted for 99.9 per cent of the variation (Fig. 2). The most important variables in separating the four samples were: NP, F, FT, and RT. An *F*-test revealed all populations to be signficantly different from one another ($P<0.001$, df=5, 67), and a classification matrix produced in the analysis revealed 100 per cent correct classification of all individuals used in this portion of the study.

Geographic Variation of Trill Calls

Means and standard deviations of variables for each group are presented in Table 1. Discriminant analysis of the 12 populations of *S. tridecemlineatus* revealed that all six call variables accounted for a signficant amount of variation exhibited by the data (*F*-test: $P<0.001$, df=11,212). The order of importance of these variables was PRR, NP, F, RT, FT, and MEP. The first four canonical variates produced by the analysis accounted for 94.2 per cent of the total dispersion.

A posterior classification matrix of each of the individual calls was produced in the study. The matrix best separated La Junta from the other populations. For that population (La Junta), nine of the 12 calls (75%) were classified correctly on the basis of the six variables used. Scott City had the poorest classification, as only three of the 20 (15%) were placed into the correct group. Only 116 of 230 (51%) of the individual calls were placed in their proper populations.

The discriminant analysis revealed that all pairs of the groups were significantly different from one another ($P<0.01$) except for the Akron— Scott City and Norton—St. Francis pairs. Fig. 3 shows the spatial relationships of the populations based on the first two canonical variates. Lines connect populations to their nearest neighbors and are based on the first

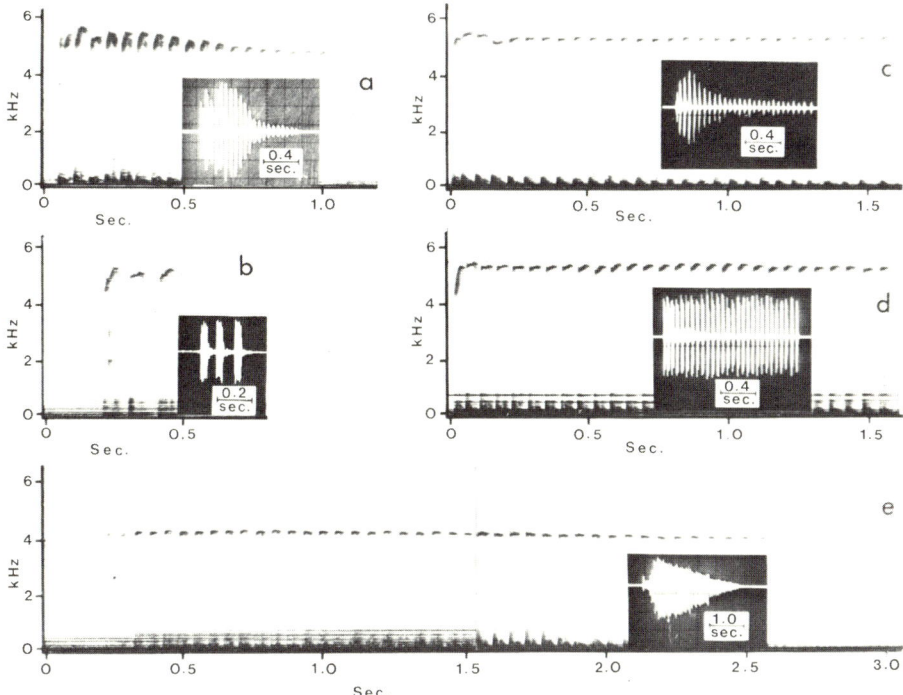

FIG. 1.—Sonagrams and oscillograms of trill vocalizations of four populations of *Spermophilus*: a, *S. tridecemlineatus* (Lubbock); b, *S. spilosoma annectens* (Island); c and d, *S. s. marginatus* (Kermit); and e, *S. mexicanus* (Monahans).

three canonical variates. Groupings based on latitude, soil types, rainfall patterns, and elevations were tested to see if the variations correlated with some spatial or environmental factor, but none was revealed. The first three canonical variables accounted for 87.4 per cent of the total variation within these 12 populations of *S. tridecemlineatus arenicola*.

DISCUSSION

Marler (1955, 1961) found calls used as warnings to be generally short, high-pitched, and without discrete beginning or end. Because of these characteristics, those calls contain few cues revealing their location and thus expose the calling animal to a minimum of danger. On the other hand, calls facilitating location generally were repetitive and of low frequency and contain most energy at their beginning. These location calls are used to establish contact with other members of the population and not as warning calls.

The trill calls produced by *Spermophilus* are often used as a warning or to alarm other members of the population (Balph and Balph, 1966; Robinson, 1980; Sherman, 1977). The trill calls of *S. tridecemlineatus, S.*

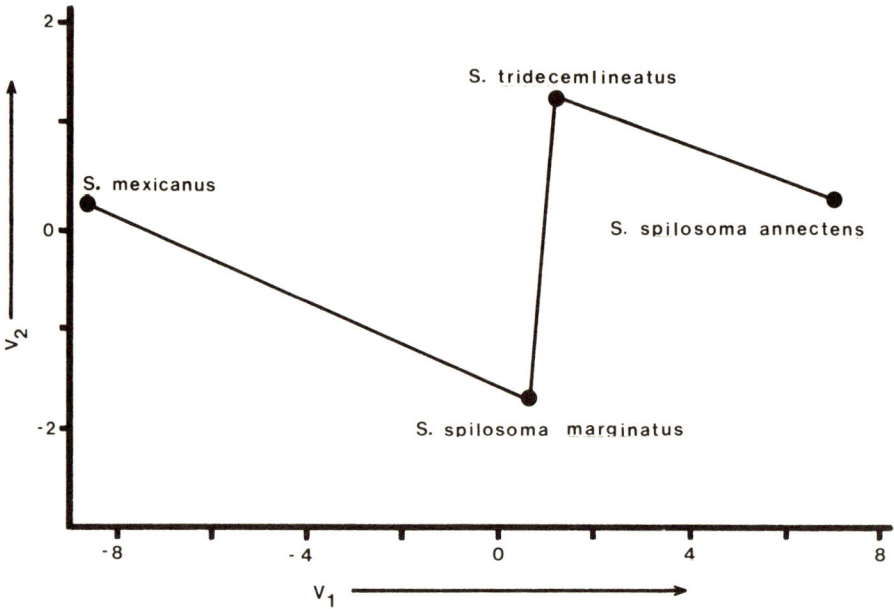

Fɪɢ. 2.— Means of trill calls of four populations of *Spermophilus* projected on the first two canonical variates (V_1 : V_2). The populations represented are: *S. mexicanus* (Monahans); *S. spilosoma marginatus* (Kermit); *S. tridecemlineatus* (Lubbock); and *S. s. annectens* (Island).

mexicanus, and *S. s. marginatus* all have calls that contain cues for location as described by Marler (1955, 1961). These species produce trill calls made up of a relatively long series of repetitive pulses, and thus individuals are easy to locate in the field (personal observation). A similar call type has been reported for *S. beldingi* (Turner, 1973; Sherman, 1977; Robinson, 1981). *Spermophilus armatus* also has an alarm call that is easy to locate, and Balph and Balph (1966) concluded that "good visual coverage of environment is their defense against predation rather than habits that are secretive or calls that are hard to locate."

Warning calls of *S. s. annectens* exhibit the characteristics described by Marler (1955, 1961). The trill call of this species is short and contains few repetitive pulses. Field observations have shown these calls to be difficult to locate. This type of call may be an adaptation to existence in an environment with limited visibility. Both the topography (sand dunes) and the vegetation tend to limit the visibility in this area. Judd *et al.* (1977) and Leonard *et al.* (1978) have described the topography and vegetation of South Padre Island and found it similar to that of Mustang Island. In addition, this species of squirrel tends to be very secretive in its behavior (personal observation). An alarm call with few cues revealing location would give the calling animal a selective advantage over one containing cues for location in habitat with limited visibility.

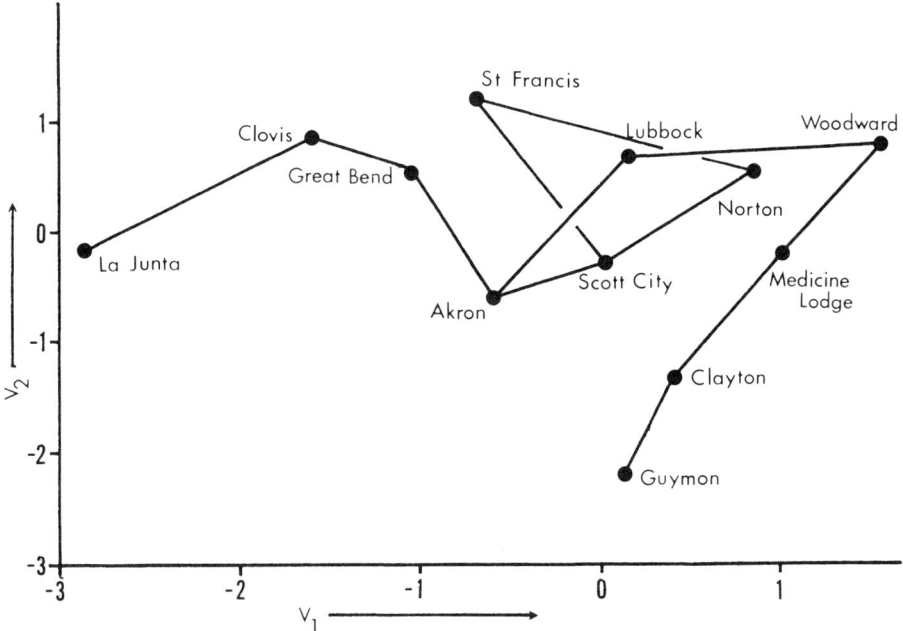

FIG. 3.—Spatial relationships of the means of 12 populations of *Spermophilus tridecemlineatus* as revealed by the discriminant analysis. The two axes represent the first two canonical variates ($V_1 : V_2$). Lines connecting populations show closest relationships.

There is a considerable amount of geographic variation of trill call variables within the range of *S. tridecemlineatus arenicola*. Stepwise discriminant analysis of these vocalizations showed statistically significant differences ($P < 0.01$) between all pairwise combinations of populations except for Akron—Scott City and Norton—St. Francis (Fig. 3). Statistical signficance is contradicted when one examines the classification matrix obtained from the same analysis. A large proportion of the individuals (49%) were incorrectly classified, even though the populations were statistically different from one another. Statistical significance obtained in the analysis may result in part from techniques in the analysis that tend to minimize the within-group variance and to maximize the between-group variance (Cooley and Lohnes, 1971).

Recent studies (LeBoeuf and Peterson, 1969; Somers, 1973) reported dialects in vocalizations in mammals other than man. Marler and Tamura (1962) restricted the use of dialect to consistent differences prevalent in the predominant call type between populations. Conner (1982) recommended that dialect be used only when interbreeding is possible between the populations being studied, as isolated populations may be subject to different evolutionary and environmental influences. All *S. tridecemlineatus arenicola* populations were geographically separated from one another by a minimum of 50 miles and interbreeding was therefore unlikely. Conner (1982) considered 13 kilometers to be enough distance to isolate prairie dog colonies. The difference between the calls in the populations of S. *tridecemlineatus* studied should probably be considered geographic variation rather than as dialects for this subspecies of ground squirrel.

The variation found to exist within populations of S. *tridecemlineatus* may be the result of genetic influences, learning, posture, or activity (Balph and Balph, 1966; Matocha, 1975). Care must be taken as the variation might not represent simply a source of error (Owings and Leger, 1980). This variation may contain information that can be used to enhance the survival of the population.

The variation in alarm calls within populations of S. *tridecemlineatus* could be selectively advantageous by aiding in individual recognition (Nottebohm, 1972; Snowdon and Cleveland, 1980). Balph and Balph (1966) found that they could recognize some individual S. *armatus* by their vocalizations. They did not, however, determine if the squirrels were able to recognize one another by their calls. Individual recognition could reveal direction of a call, especially important to females, which tend to inhabit definite home ranges (McCarley, 1966; Sherman, 1977). This would permit rapid determination of the direction of the calling animal and potential danger.

In addition to individual recognition, variations in call types recently have been demonstrated to be used to convey behavioral messages. Owings and Leger (1980) found the chatter calls of S. *beecheyi* to be information rich, and six different variations of this call type were used by S. *beecheyi* to convey information regarding agonistic chases and encounters with mammals, snakes, and raptors. Although not revealed by my study, it may be found that the variation in S. *tridecemlineatus* trill vocalizations contains information selectively advantageous to this species of ground squirrel.

ACKNOWLEDGMENTS

I am grateful to Dr. Robert L. Packard for guidance throughout this study and for aid in the preparation of an earlier version of the manuscript. Drs. Robert J. Baker, Eric G. Bolen, John S. Mecham and Francis L. Rose offered advice on methods of research and critically read the manuscript. Some equipment was provided by Dr. J. T. Peacock of Texas A&I University, Kingsville. Drs. William R. Atchley and Duane Anderson aided in the statistical design interpretation; Mr. Bill Wyatt aided in computer procedure. Dr. Robert E. Martin assisted in collection of animals and advised throughout the study. Many fellow graduate students provided suggestions and also were helpful. Dr. John S. Mecham provided equipment for call analysis and supervised in the use of this equipment. My wife, Ruth, aided in data collection and preparation of the manuscript.

This study was financed, in part, by a grant from the Society of Sigma Xi. Certain data also were obtained when I was a research assistant in the

IBP Grasslands Biome studies supported by a grant from the National Science Foundation to Colorado State University.

LITERATURE CITED

BALPH, D. M., AND D. F. BALPH. 1966. Sound communication of Unita ground squirrels. J. Mamm., 47:440-450.

BETTS, B. J. 1976. Behaviour in a population of Columbian ground squirrels, *Spermophilus columbianus columbianus*. Anim. Behav., 25:221-230.

CONNER, D. A. 1982. Dialects versus geographic variation in mammalian vocalizations. Anim. Behav., 30:297-298.

COOLEY, W. W., AND P. R. LOHNES. 1971. Multivariate data analysis. John Wiley and Sons, New York, xii+364 pp.

DIXON, W. J. 1971. Biomedical computer programs. Univ. California, Publ. Automatic Computation, 2:x+600.

HALL, E. R. 1981. The mammals of North America. 2nd ed. John Wiley & Sons, New York, 1:xviii+600+*90*.

HOWELL, A. H. 1938. Revision of the North American ground squirrels, with a classification of the North American Sciuridae. N. Amer. Fauna, 56:1-256.

JUDD, F. W., R. I. LONARD, AND S. L. SIDES. 1977. The vegetation of South Padre Island, Texas in relation to topography. Southwestern Nat., 22:31-48.

LEBOEUF, B. J., AND R. S. PETERSON. 1969. Dialects in elephant seals. Science, 166:1654-1656.

LONARD, R. I., F. W. JUDD, AND S. L. SIDES. 1978. Annotated checklist of the flowing plants of South Padre Island, Texas. Southwestern Nat., 23:497-509.

MARLER, P. 1955. Characteristics of some animal calls. Nature, London, 176:6-8.

———. 1961. The logical analysis of animal communication. J. Theor. Biol., 1:295-317.

MARLER, P., AND M. TAMURA. 1962. Song "dialects" in three populations of white-crowned sparrows. Condor, 64:368-377.

MATOCHA, K. G. 1975. Vocal communication in ground squirrels, genus *Spermophilus*. Unpublished Ph.D. dissertion, Texas Tech Univ. 67 pp.

———. 1977. The vocal repertoire of *Spermophilus tridecemlineatus*. Amer. Midland Nat., 98:482-487.

MCCARLEY, W. H. 1966. Annual cycle, population dynamics and adaptive behavior of *Citellus tridecemlineatus*. J. Mamm., 47:294-316.

NOTTEBOHM, F. 1972. The origins of vocal learning. Amer. Nat., 106:116-140.

OWINGS, D. H., AND D. W. LEGER. 1980. Chatter vocalizations of California ground squirrels: predator—and social—role specificity. Z. Tierpsychol., 54:163-184.

OWINGS, D. H., M. BORCHERT, AND R. VIRGINIA. 1977. The behaviour of California ground squirrels. Anim. Behav., 25:221-230.

ROBINSON, S. R. 1981. Alarm communication in Belding's ground squirrels. Z. Tierpsychol., 56:150-168.

SHERMAN, P. W. 1977. Nepotism and the evolution of alarm calls. Science, 197:1246-1253.

———. 1980. The limits of ground squirrel nepotism. Pp. 505-544, *in* Sociobiology: beyond nature/nurture? (G. W. Barlow, ed.). Westview Press, Colorado.

SHIELDS, W. M. 1980. Ground squirrel alarm calls: nepotism or parental care? Amer. Nat., 116:599-603.

SNOWDON, C. T., AND J. CLEVELAND. 1980. Individual recognition of contact calls by pigmy marmosets. Anim. Behav., 28:717-727.

SOMERS, P. 1973. Dialects in southern Rocky Mountain pikas, *Ochotona princeps* (Lagomorpha). Anim. Behav., 21:124-137.

TURNER, L. W. 1973. Vocal and escape responses of *Spermophilus beldingi* to predators. J. Mamm., 54:990-993.

HABITAT RELATIONSHIPS OF FOUR SPECIES
OF MICE IN SOUTHWESTERN ARKANSAS

JAMES B. MONTGOMERY, JR.

Throughout much of North America several species of morphologically similar rodents often appear to coexist but, upon closer examination, are found to occupy different microhabitats. This is the apparent situation in the Ouachita Mountains of Arkansas and Oklahoma where the Texas (brush) mouse (*Peromyscus attwateri*), cotton mouse (*Peromyscus gossypinus*), white-footed mouse (*Peromyscus leucopus*), and the golden mouse (*Ochrotomys nuttalli*) occur in forest habitats. Numerous studies of the habitat relationships of various combinations of those mice in sympatry have been published, including studies of *P. attwateri* with *P. leucopus* (Blair, 1938; Brown, 1964; Long, 1961), *P. leucopus* with *P. gossypinus* (McCarley, 1954, 1963), *P. leucopus* with *O. nuttalli* (Dueser and Shugart, 1978, 1979), and *P. gossypinus* with *O. nuttalli* (McCarley, 1958; Packard, 1968). Except for data published in checklist format (Blair, 1939; Sealander, 1956, 1979), habitat relationships of those mice have not been reported from the Ouachita Mountains region where all four species occur together. This paper is a report of the habitat relationships of those four species in a forested area adjacent to the Cossatot River in southwestern Arkansas.

This study was conducted in northwestern Howard County, Arkansas, within a tract of forest from the Highway 4 bridge over the Cossatot River to the confluence of Opossum Creek with the river 6 kilometers to the southwest. Lower portions of the study area now lie within the shoreline of Gillham Reservoir. Vegetation of that area is primarily of the pine-oak-hickory forest type (Moore, 1972). In Howard County, the Cossatot River flows southward within a narrow entrenched valley across numerous parallel east-west ridges separated by small narrow valleys (Miser and Purdue, 1929). The structure of the forest is strongly influenced by this topography. When viewed from the air or from aerial photographs, for example, forest on north and east facing slopes differs in appearance from forest on south and west facing slopes, primarily in the degree of canopy cover and the ratio of shortleaf pine (*Pinus echinata*) to hardwood trees. This topographic pattern may be divided, for comparative purposes, into categories of upland, north (and east) facing slopes, south (and west) facing slopes, and lowland. Superimposed on the topographic pattern are changes in the structure and composition of the forest resulting from past logging operations.

METHODS

Mice were captured with Museum Special snaptraps baited with peanut butter and placed approximately 10 meters apart in a trapline consisting of 25 to 50 stations. Additional traplines were set parallel to the first with distances of 25 to 30 meters between lines. These traplines were oriented at a right angle to the major local terrain feature, such as a ridge or stream, so as to cross parallel bands of differing habitats. Thirty localities within the 10-square kilometer study area were trapped with 77 traplines (2.6 per locality). Traps were set for an average of 1.3 nights from January through April and for 2.6 nights from May to July for a total of 4670 trapnights between 23 January and 8 July 1972. That period encompassed a spring peak in rodent activity, with the majority of captures occurring from March through May. Although an effort initially was made to select randomly trapping localities, localities were deliberately chosen during the latter part of the trapping period in order to obtain a larger sample of *Peromyscus leucopus* capture sites. That resulted in undersampling some habitats and oversampling others.

Sketch maps showing locations of traps and captures in relation to the terrain were prepared for each trapping area, and capture sites were numbered and marked for subsequent quantification of vegetation cover. In order to increase sample sizes, capture sites of those subadult and juvenile mice for which species identity unquestionably could be determined were included in the habitat comparisons. No differences between adults and younger mice of each species (or between sexes) were found ($P<0.05$) by univariate analysis of variance tests of eight variables describing habitat structure of the capture sites.

Spatial overlap of species pairs was estimated from numbers of captures in upland, north (and east) slope, south (and west) slope, and lowland habitats. Capture sites were assigned to one of the four categories and trapping effort (number of trapnights in each category) was determined from field notes and and the sketch maps. Actual numbers of captures were converted to numbers of captures per 1000 trapnights and indices of overlap were calculated by Horn's (1966) formula:

$$Ro = \frac{(x_i + y_i)\log(x_i + y_i) - x_i\log x_i - y_i\log y_i}{(X + Y)\log(X + Y) - X \log X - Y \log Y}$$

Eight variables (Table 1) estimating overstory, understory, and ground cover components of habitat structure were measured at the capture sites during July and August 1972 by methods decribed by James and Shugart (1970) and used by James (1971). Tree basal area was determined from diameter estimates of all trees (woody plants greater than 7.6 centimeters in diameter) within a 0.02 hectare circular plot (a radius of 8 meters) centered on the capture site, and the height of the tallest tree within the plot was determined by trigonometry. Shrub stems (woody plants taller than 1 meter

TABLE 1.—*Results of one-way analyses of variance tests (F) and means ($\bar{X}$) of each species on eight variables describing habitat structure. Means labeled with the letter "a" within a row are not different at the 95% level of probability. Means were calculated from untransformed data. See text for description of the variables. (**=P<0.01 and ***=P<0.001).*

Variable	Analysis of variance (df=3, 208) F	Peromyscus attwateri (N=109) $\bar{X}$	Peromyscus gossypinus (N=32) $\bar{X}$	Peromyscus leucopus (N=25) $\bar{X}$	Ochrotomys nuttalli (N=46) $\bar{X}$
Overstory					
Tree basal area (in square meters)	5.370**	0.44	0.56ᵃ	0.56ᵃ	0.51ᵃ
Height of tallest tree (in meters)	16.053***	14.5	18.8ᵃ	17.3ᵃ	18.1ᵃ
% Canopy cover	26.139***	67.1	86.3ᵃ	85.3ᵃ	83.3ᵃ
Understory					
Number of shrub stems	10.523***	37.1ᵃ	29.2ᵃ	21.5	34.5ᵃ
Number of vine stems	15.188***	5.3	31.8ᵃ	0.9	18.7ᵃ
Ground cover					
% Leaf litter	16.197***	51.6	58.0ᵃ	72.0	60.2ᵃ
% Ground vegetation	8.509***	37.0ᵃ	36.8ᵃ	26.3	37.9ᵃ
% Exposed rock	32.045***	7.4	2.2ᵃ	0.6ᵃ	0.3ᵃ

and less than 7.6 centimeters in diameter) and vine stems encountered higher than 1 meter above ground were counted within two 20 square meters armslength rectangles oriented along north-south and east-west axes across the center of the plot. Canopy cover and ground cover were estimated in percentages by the ocular tube method (James and Shugart, 1970). Fifteen skyward and downward sightings were taken at 0.9 meter intervals along the center of each armslength rectangle. Percentages were determined from the 30 total sightings using the criterion of presence or absence as viewed at the intersection of the cross threads in the ocular tube. Categories of ground cover were ground vegetation (all grasses, forbs, seedlings, and recumbent vines), leaf litter (both leaves and pine needles), and exposed rock.

As noted, vegetation structure of the capture sites was not quantified until up to six months after the mice were captured. Several of the variables, mainly canopy cover and ground vegetation, exhibited seasonal changes in appearance during this period but were nevertheless included in the habitat analysis. Their inclusion would be invalid if habitat use by the mice varied seasonally. No seasonal habitat shifts, however, were noticed; the species of mice were trapped in the same respective habitat types during late spring and summer as during late winter and early spring. Secondly, the variables were not selected to represent specific factors used as cues by forest rodents but were chosen to provide a basis for comparing distributions of the mice in relation to overall forest structure. Vegetation was measured during mid to late summer, rather than throughout the trapping period, in order to avoid the effects of seasonal variation in those variables when making habitat comparisons.

A total of 212 samples representing capture sites of 109 *Peromyscus attwateri*, 32 *P. gossypinus*, 25 *P. leucopus*, and 46 *Ochrotomys nuttalli* were used in the statistical analyses. For situations where two mice were captured at the same site, the same data were used for each individual; 21 samples were thus duplicated. One-way analysis of variance with logarithmic data transformations and the Student-Newman-Keul's procedure to identify homogeneous subsets were used to determine the ability of each variable to separate species (Table 1). Principal coordinates analysis of the eight variables (standardized so that the mean equals zero and variance equals one and then transformed into an Euclidean distance matrix) was used to ordinate the species of mice along axes representing major components of variation in habitat structure (Gower, 1966; James, 1971; Rohlf, 1972). Photographs of capture sites and field notes were used as aids in interpreting these axes. Ninety-five per cent confidence ellipses around bivariate species means in the space defined by the first two coordinate axes were calculated to show species separation (Dueser and Shugart, 1979). The principal coordinates analysis was computed by an IBM 370 computer at Texas Tech University, with the NT-SYS package of computer programs (Rohlf *et al.* 1972). All other procedures were calculated with an electronic calculator by methods described in Sokal and Rohlf (1969).

RESULTS

Comparisons of captures by topographic category (Table 2) show that all species were nonrandomly distributed relative to topography. *Peromyscus* distributions were characterized by inverse spatial densities, with each species occurring in greatest abundance in a different topographic setting. Indices of abundance calculated from the capture data show relatively little overlap between the species of *Peromyscus* (indices are 0.25 for *P. attwateri* with *P. leucopus*, 0.37 for *P. attwateri* with *P. gossypinus*, and 0.40 for *P. gossypinus* with *P. leucopus*). Overlap between genera, on the other hand, was greater (index of overlap, 0.86) with the distribution of *Ochrotomys nuttalli* being most similar to that of *P. gossypinus* (index of overlap, 0.94). Overlaps between *O. nuttalli* and the other species of *Peromyscus* were less (0.55 with *P. attwateri* and 0.58 with *P. leucopus*) but still greater than overlaps between the species of *Peromyscus*.

Habitat relationships comparable to those shown by the indices of overlap and capture data also are evident in the positions of the species means along the first two axes of the principal coordinates analysis of habitat structure (Fig. 1). Ecological interpretations of those axes can be derived from the relative associations of individual variables with each coordinate axis (Table 3). For example, variables related to large trees (leaf litter, canopy cover, tree height, and tree basal area) have large negative associations, and variables representing brushy woodland (shrub and vine stems, ground vegetation, and exposed rock) have large positive associations with the first axis.

TABLE 2.—*Distribution of captures in relation to topography. Numbers of captures in each topographic division (projected numbers of captures per 1000 trapnights are listed in parentheses) and chi-square tests of independence of species captures and topography (**=P<0.01).*

Topographic division.	Number of trap nights.	Peromyscus attwateri	Peromyscus gossypinus	Peromyscus leucopus	Ochrotomys nuttalli
Upland	1685	3 (1.8)	6 (3.6)	23 (13.6)	13 (7.7)
North & East Facing Slopes	830	1 (1.2)	6 (7.2)	1 (1.2)	8 (9.6)
South & West Facing Slopes	1395	116 (83.2)	1 (0.7)	1 (0.7)	9 (6.5)
Lowland	760	6 (7.9)	22 (28.9)	0 (0.0)	17 (22.4)
Calculated χ^2		233.51**	58.71**	34.00**	14.18**

Photographs of capture sites with large negative scores along this axis show a forest of large tall trees with a closed canopy, few understory trees or saplings, and very little ground vegetation. Conversely, those capture sites with large positive scores were in brushy woodland consisting of saplings and small trees with grasses, seedlings, and forbs growing in open areas between scattered trees. Some of these sites were on steep rocky slopes where vegetation reflects exposure and soil conditions and other sites were in areas where all marketable timber had apparently been removed. This axis, which accounted for 36.7 per cent of the total variation in the analysis, represents a continuum from closed canopy forest to open brushy woodland. Means and standard deviations of the species of mice along this axis are -0.60 ± 0.506 for *Peromyscus leucopus*, -0.18 ± 0.485 for *P. gossypinus*, -0.15 ± 0.575 for *Ochrotomys nuttalli*, and 0.25 ± 0.454 for *P. attwateri*.

Understory and ground cover variables are most strongly associated with the second coordinate axis. Canopy cover and tree basal area, on the other hand, have more neutral associations, suggesting that forest overstory does not vary greatly along the gradient represented by this axis. The combination of tall trees, numerous vine stems, and a large amount of ground vegetation (large positive associations) was most often encountered in lowland vegetation. Field data and photographs show that all sites with high positive scores on the second axis were either on the river bank, in forest on low lying land near the river, or in forest or forest edge along small intermittent streams. Shrub stems and leaf litter indicate brushy understory under larger trees and exposed rock suggests slopes and ridges. As shown by photographs and field notes, sites with large negative scores were on steep slopes and in upland areas characterized by rock outcrops and a brushy understory. This axis, which accounted for 18.4 per cent of the total variation, represents a gradient of decreasing woody understory vegetation and increasing vines and ground vegetation associated with the transition from rocky or upland sites to lowland sites. Means and standard deviations of the species along this axis are -0.17 ± 0.339 for *P. attwateri*,

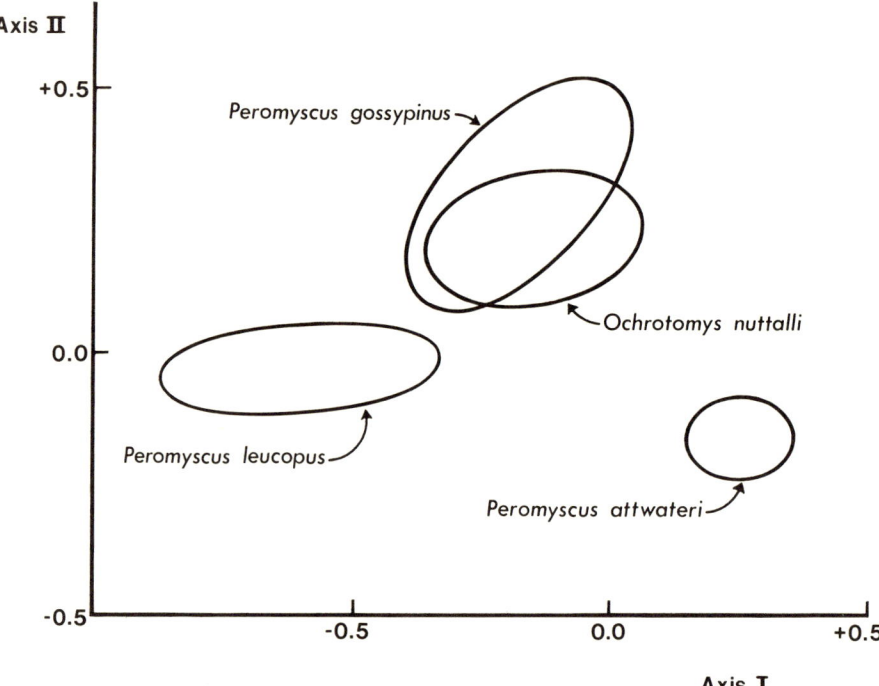

Fig.1.—Ordination of *Peromyscus* and *Ochrotomys* by principal coordinates analysis of capture site habitat structure: 95 per cent confidence ellipses around bivariate species means on the first two principal coordinate axes. Interpretations of the axes are discussed in the text.

−0.03±0.168 for *P. leucopus*, 0.21±0.361 for *Ochrotomys nuttalli*, and 0.30±0.490 for *P. gossypinus*.

Each of the eight habitat variables had significant ability ($P<0.01$) to separate at least one species of *Peromyscus* from the other species of mice (Table 1) but each alone contributed only 12.5 per cent of the total available habitat information. The species ordination by principal coordinates analysis showed habitat relationships and demonstrated separation of the three species of *Peromyscus* along gradients of habitat structure that together accounted for 55.1 per cent of the available information. Habitat structure measured in the vicinity of captures, however, was not sufficient to show any ecological separation between *P. gossypinus* and *Ochrotomys nuttalli*.

Additional information relevant to the distribution of *Orchotomys nuttalli* is contributed by the third coordinate axis, which accounted for 16.6 per cent of the total variation. Capture sites with high positive scores along this axis, regardless of their locations relative to topography, were covered with dense stands of saplings and shrubs (mostly shortleaf pine and hardwood species that comprise the overstory) and small trees (7.6 to 15 centimeters in diameter), with few larger trees. Sites with large negative scores were typically in the vicinity of rock outcrops in south slope forest

TABLE 3.—*Summary of principal coordinates of vegetation structure at* Peromyscus *and* Ochrotomys *capture sites: identification of coordinate axes.*

	Principal coordinate axes		
	I	II	III
Per cent of total variation accounted for by each axis	36.7	18.4	16.6
Cumulative percentage		55.1	71.7
Associations of variables with coordinate axes			
Tree basal area	−5.717	2.889	−4.765
Height of tallest tree	−6.192	4.810	−1.535
% Canopy cover	−8.738	1.056	−0.409
Number of shrub stems	8.379	−5.540	9.535
Number of vine stems	4.249	6.290	1.408
% Leaf Litter	−10.823	−5.765	5.701
% Ground vegetation	12.381	6.249	−0.051
% Exposed rock	6.462	−9.989	−9.883

characterized by larger trees with fewer saplings or shrubs in the understory. The relationship between the ends of this axis is obscure. Nevertheless, this axis may be interpreted as a gradient of increasing tree reproduction in response to increasingly severe disturbance, primarily by past logging operations. Means and standard deviations of the mice on this axis are −0.07±0.445 for *Peromyscus attwateri*, −0.02±0.341 for *P. gossypinus*, 0.05±0.216 for *P. leucopus*, and 0.15±0.267 for *O. nuttalli*. The only significant separation between species, as shown by nonoverlapping 95 per cent confidence limits, is between *P. attwateri* and *O. nuttalli*. The upper confidence limit of *P. attwateri* is 0.02 and the lower confidence limit of *O. nuttalli* is 0.06. All other combinations of species confidence limits overlap on the third axis and the confidence limits of all four species overlap on the fourth and subsequent axes.

DISCUSSION

The comparisons of capture sites of the three species of *Peromyscus* within the forest along the Cossatot River demonstrated microhabitat separation associated with vegetation structure. This type of spatial separation of similar species is consistent with published reports of ecological relationships between species of mice in other areas of North America (for example: Brown, 1964; Dueser and Shugart, 1978, 1979; Geluso, 1971; Holbrook, 1978; M'Closkey and Fieldwick, 1975). The lowland-upland relationship between *P. gossypinus* and *P. leucopus* described in this study is well documented (Calhoun, 1941; Dice, 1940; McCarley, 1954, 1963) and the brushy woodland versus closed canopy forest relationship between *P. attwateri* and *P. leucopus* has likewise been reported (Brown, 1964; Long, 1961).

The habitat of *Ochrotomys nuttalli* along the Cossatot River also agrees with published habitat descriptions (such as those cited in the introduction). This species seemed to be relatively common, with captures distributed throughout the study area. Most of the captures, however, were in lowland habitat also occupied by *P. gossypinus*. In contrast, Dueser and Shugart (1978, 1979) found *O. nuttalli* to be relatively uncommon (compared to *P. leucopus*) in the hardwood forest of their eastern Tennessee study area. That difference probably is related to vegetation type. Dueser and Shugart described *O. nuttalli* as a habitat specialist with a preference for an evergreen (pine) canopy, an uncommon component of the vegetation in their study area. That vegetation type, however, is widespread along the Cossatot River and, if a preference for pine trees is characteristic of *O. nuttalli*, then the greater abundance and local distribution of this species would be expected.

Indices of overlap calculated from the capture data revealed greater habitat overlap between *Ochrotomys nuttalli* and the species of *Peromyscus* than between the three species of *Peromyscus*. Interestingly, an index of overlap between *O. nuttalli* and *P. gossypinus* in eastern Texas, calculated from capture data published by Packard (1968), is similar to the index of overlap between *Ochrotomys* and *Peromyscus* in this study (0.87 and 0.86, respectively). The greater intergeneric overlap could indicate less competition between *Ochrotomys* and *Peromyscus* for space or habitat or might simply result from an omission of another, possibly critical, variable such as "use of arboreal vegetation."

Holbrook (1979) concluded that difference in arboreal activity is an important factor allowing coexistence of cricetine rodents, and Montgomery (1979) presented data that suggests differential use of vegetation above ground enables the coexistence of *P. attwateri* and *P. leucopus* in riparian vegetation in Texas. McCracken (1978) reported differences between *Ochrotomys* and *Peromyscus* in arboreal activity in an artificial tree and concluded that *O. nuttalli* is dominant over *P. gossypinus* and *P. leucopus* in an arboreal setting. In the present study, the separation of *O. nuttalli* and *P. attwateri* along the third principal coordiante axis, which represents a type of brushyness, suggests a vertical (or arboreal) component in the relationship between these two species. These observations suggest that differences in vegetation use above ground could allow the apparent coexistence of *O. nuttalli* with *P. gossypinus* and, to a lesser extent, with the other species of *Peromyscus*, but no conclusion can be drawn from the available data.

ACKNOWLEDGMENTS

This report is based on a thesis submitted in 1974 in partial fulfillment of the requirements for the degree of Master of Science at the University of Arkansas, Fayetteville. I wish to thank the members of my advisory committee, Drs. John A. Sealander (Chairman), Edward E. Dale, Douglas

A. James, and James M. Walker, for their assistance and guidance during the course of this study. Drs. Emmet T. Hooper and Robert L. Packard examined and identified some of the mice. Traps were purchased, and a portion of the travel expenses was defrayed, by the Department of Zoology, University of Arkansas. J. Waid Griffin, Jr. prepared the figure. I also wish to thank the anonymous reviewers of an earlier draft of this paper.

LITERATURE CITED

BLAIR, W. F. 1938. Ecological relationships of mammals of the Bird Creek region, northeastern Oklahoma. Amer. Midland Nat., 20:473-526.

———. 1939. Faunal relationships and geographical distribution of mammals in Oklahoma. Amer. Midland Nat., 22:85-133.

BROWN, L. N. 1964. Ecology of three species of *Peromyscus* from southern Missouri. J. Mamm., 45:189-202.

CALHOUN, J. B. 1941. Distribution and food habits of mammals in the vicinity of the Reelfoot Lake Biological Station, II. Discussion of the mammals recorded from the area. J. Tenn. Acad. Sci., 16:207-225.

DICE, L. R. 1940. Relationships between the wood-mouse and the cotton-mouse in eastern Virginia. J. Mamm., 21:14-23.

DUESER, R. D., AND H. H. SHUGART, JR. 1978. Microhabitats in a forest-floor small mammal fauna. Ecology, 59:89-98.

———. 1979. Niche pattern in a forest-floor small mammal fauna. Ecology, 60:108-118.

GELUSO, K. N. 1971. Habitat distribution of *Peromyscus* in the Black Mesa region of Oklahoma. J. Mamm., 52:605-607.

GOWER, J. C. 1966. Some distance properties of latent root and vector methods in multivariate analysis. Biometrika, 53:325-338.

HOLBROOK, S. J. 1978. Habitat relationships and coexistence of four sympatric species of *Peromyscus* in northwestern New Mexico. J. Mamm., 59:18-26.

———. 1979. Habitat utilization, competitive interactions, and coexistence of three species of cricetine rodents in east-central Arizona. Ecology, 60:758-769.

HORN, H. S. 1966. Measurement of "overlap" in comparative ecological studies. Amer. Nat., 100:419-424.

JAMES, F. C. 1971. Ordinations of habitat relationships among breeding birds. Wilson Bull., 83:215-236.

JAMES, F. C., AND H. H. SHUGART, JR. 1970. A quantitative method of habitat description. Audubon Field Notes, 24:727-736.

LONG, C. A. 1961. Natural history of the brush mouse (*Peromyscus boylii*) in Kansas with a description of a new subspecies. Univ. Kansas Publ., Mus. Nat. Hist., 14:99-110.

MCCARLEY, W. H. 1954. The ecological distribution of the *Peromyscus leucopus* species group in eastern Texas. Ecology, 35:375-379.

———. 1958. Ecology, behavior and population dynamics of *Peromycus nuttalli* in eastern Texas. Texas J. Sci., 10:147-171.

———. 1963. Distributional relationships of sympatric populations of *Peromyscus leucopus* and *P. gossypinus*. Ecology, 44:784-788.

MCCRACKEN, D. W. 1978. A study of utilization and partitioning of vertical space in five species of small, woodland rodents (*Peromyscus leucopus, Peromyscus maniculatus, Peromyscus gossypinus, Ochrotomys nuttalli*, and *Clethrionomys gapperi*). Unpublished Ph.D. dissertation, Wake Forest University, Winston-Salem, North Carolina, 126 pp.

M'CLOSKEY, R. T., AND B. FIELDWICK. 1975. Ecological separation of sympatric rodents (*Peromyscus* and *Microtus*). J. Mamm., 56:119-129.

MISER, H. D., AND A. H. PURDUE. 1929. Geology of the De Queen and Caddo Gap quadrangles, Arkansas. Bull. U.S. Geol. Surv., 808:xi+1-195.

MONTGOMERY, J. B., JR. 1979. Behavioral interactions and local distribution of three species of *Peromyscus* in South Ceta Canyon, Randall County, Texas. Unpublished Ph.D. dissertation, Texas Tech University, Lubbock, Texas, vii+130 pp.

MOORE, D. M. 1972. Trees of Arkansas. Arkansas Forestry Commission, Little Rock, Arkansas, 142 pp.

PACKARD, R. L. 1968. An ecological study of the fulvous harvest mouse in eastern Texas. Amer. Midland Nat., 79:68-88.

ROHLF, F. J. 1972. An empirical comparison of three ordination techniques in numerical taxonomy. Syst. Zool., 21:271-280.

ROHLF, F. J., J. KISHPAUGH, AND D. KIRK. 1972. Numerical taxonomy system of multivariate statistical programs. Version of 1972. State Univ. of New York, Stony Brook.

SEALANDER, J. A., JR. 1956. A provisional check-list and key to the mammals of Arkansas (with annotations). Amer. Midland Nat., 56:257-296.

———— . 1979. A guide to Arkansas mammals. River Road Press, Conway, Arkansas, x+313 pp.

SOKAL, R. R., AND F. J. ROHLF. 1969. Biometry. W. H. Freeman and Company, San Francisco, xiii+776 pp.